Praise for Tapan Munroe's
What Makes Silicon Valley Tick?
The Ecology of Innovation at Work

"What Makes Silicon Valley Tick offers a fascinating, up-to-date account of Silicon Valley's enduring vitality. Anyone interested in understanding the region that continues to invent the future should read this insightful book."

AnnaLee Saxenian, Dean and Professor, School of Information,
University of California, Berkeley

"What Makes Silicon Valley Tick is a tour de force: comprehensive, systematic, authoritative, and highly readable. It showcases what Silicon Valley insiders have known for a long time—that Tapan Munroe is our foremost interpreter and expositor."

Russell Hancock, President & Chief Executive Officer,
Joint Venture: Silicon Valley Network

"Tapan Munroe admirably captures the spirit of Silicon Valley—both its working parts and the ethos that drives it. What comes across clearly is its unplanned, organic nature, which makes the Silicon Valley experience hard to replicate. As Munroe points out, its story in many ways describes the entire San Francisco Bay Area, and explains why the region has emerged as a center of global innovation. While, as Munroe candidly points out, it faces significant challenges, the dynamism and flexibility of the Silicon Valley model holds an important key to U.S. competitiveness, where the ability to innovate and create new value will differentiate communities that will prosper from those that will not in the future."

Dr. R. Sean Randolph, President & CEO,
Bay Area Council Economic Institute

Praise for Tapan Munroe's
What Makes Silicon Valley Tick?
The Ecology of Innovation at Work
(Continued)

"This book does a terrific job of analyzing the innovation economy of Silicon Valley. Understanding the lessons from the Valley's success at this time of global crisis is of paramount importance to ensure innovation and economic prosperity. It is a must read for leaders of business, venture capitalists, universities and research organizations who would like to replicate the innovative magic of Silicon Valley in their own region."

Richard C. Atkinson, President Emeritus, University of California

"A superbly written book, it demonstrates the authors' insight into the Silicon Valley technology enterprise and the innovation process. In addition to entrepreneurs and managers of technology, I would highly recommend this book to engineering and business schools and anyone interested in the management of technology and the innovation process."

Ravi Jain, PhD, PE, Dean and Professor, School of Engineering and Computer Science, University of the Pacific

"If ever we needed to understand and develop innovation economies globally, the time is now, in the face of the worst economic crisis since the Great Depression. Tapan Munroe's highly engaging and penetrating analysis is a must read for leaders and entrepreneurs across the private, public, and nonprofit sectors. Munroe deconstructs the region that still leads the world in technology. He proposes a timely model to apply in meeting both anticipated and unanticipated global economic challenges. This book is an inspiration and incentive for us all to develop and sustain transformational ecologies of innovation as an antidote for a seriously ailing global economy."

Barbara K. Bundy, PhD, Executive Director, Center for the Pacific Rim, University of San Francisco

"Tapan Munroe lucidly explains why this extraordinary innovation economy continues to reinvent itself time and again despite many challenges for more than a half a century. Silicon Valley's dynamic innovation ecosystem, as described and analyzed by the authors, provides various Science Parks and technology regions around the world a powerful model to emulate and a benchmark to assess their performance. This is a must read for all who are interested in understanding the key to a successful and sustainable economy in the twenty-first century."

Dr. Felipe Romera Lubias, President, Association of Science and Technology Parks of Spain (APTE)

"In this clear, compelling volume, Tapan Munroe has distilled the very essence of Silicon Valley. As the prototypical cluster of innovation, Silicon Valley is often held up as a unique ecosystem. Munroe deconstructs its uniqueness and helps us understand it, one key element at a time. Anchored by a strong, deep understanding that comes from decades of experience in the Valley, Munroe's profoundly clear analysis makes a very helpful guide for anyone wanting to navigate or do business with this resilient entrepreneurial community."

Jerome S. Engel, Executive Director, Lester Center for Entrepreneurship & Innovation, University of California, Berkeley

"Tapan highlights what can and must be done if the U.S. is to remain a viable and productive economy that counts in world economic affairs. He is to the East Bay what Thomas Friedman is to the East Coast. Tapan documents and identifies the megatrends that will guide economic and social development in this century."

Douglas W. Borchert, Northern California Underwriting Counsel, Fidelity National Title Insurance Company

"In times of economic change and stress, there is no better tonic than Tapan Munroe's compelling vision of a business environment driven by creative innovation. This book, so excellently researched and written, provides a compelling basis for confidence in what can be accomplished if we choose to learn from the remarkable culture that is Silicon Valley. In an age when so much seems to have been lost due the unrestrained greed of a few, his hopeful perspective provides a foundation for collaboration which can benefit us all."

Rev. Peter Whitelock, Senior Pastor, Lafayette Orinda Presbyterian Church, Orinda, California

"In a world where so many aspects of our lives are globalized, Silicon Valley casts a long shadow both as a center and symbol of innovation. And in a world that is currently mired in a serious recession, *What Makes Silicon Valley Tick?* is a lively and well-illustrated read for anyone interested in the economics of hope. Choosing a biological metaphor, Munroe traces the growth and continuing evolution of Silicon Valley by aptly identifying and describing the seven factors crucial to its success. In doing so, he not only educates us about the importance of innovation in today's economy, he provides us with hope for the future."

Thomas Scovel, Professor of Linguistics, San Francisco State University

"A long-time observer of the region, Dr. Munroe offers an insightful view of Silicon Valley's remarkable evolution. Like a learning organism, the Valley is genetically driven to thrive in the face of relentless change, constantly regenerating and diversifying its relationship with the world around it. As the preeminent innovation-driven economy, Silicon Valley may offer the best proof yet that innovation is not a luxury, but a necessity, and the best and only way to sustain prosperity."

Robert Brant, Executive Vice President, City National Bank, San Francisco and Beverly Hills

What makes
Silicon Valley tick?

The Ecology of Innovation at Work

Tapan Munroe, PhD

with Mark Westwind, MPA

Available from Nova Vista Publishing

Business Books
Win-Win Selling
Vendre gagnant gagnant (French Version of Win-Win Selling)
Versatile Selling
S'adapter pour mieux vendre (French Version of Versatile Selling)
Social Styles Handbook
I Just Love My Job!
Taking Charge of Your Career Workbook
Leading Innovation
Grown-Up Leadership
Grown-Up Leadership Workbook
Time Out for Leaders
Service Excellence @ Novell
What Makes Silicon Valley Tick?

Nature Books
Return of the Wolf
The Whitetail Fieldbook

Music Books
Let Your Music Soar

How to order: single copies may be ordered from www.novavistapub.com. In North America, you may call +1-503-548-7597. Elsewhere, you may call +32-476-360-989. Please contact info@novavistapub.com for special sales and bulk quantity discounts.

The map on page 18 is reprinted with the permission of Silicon Valley Map. Please visit www.siliconvalleymap.com for more information.

ISBN 978-90-77256-28-2

Cover and text design: Annette Krammer, Forty-two Pacific, Inc.

Editorial: Molly Kelley

Printed in the United States of America

20 19 18 17 16 15 14 13 12 11 10 9 8 7 6 5 4 3 2

Never before in history has innovation offered the promise of so much to so many in so short a time.

Bill Gates
Quoted on ThinkExist.com

Contents

Innovation is the central issue in economic prosperity.

Michael Porter, Harvard University

Today the U.S. is in the midst of one of the worst economic downturns since the Great Depression of the 1930s. The housing debacle and credit crisis have decimated the housing and financial industries. Our domestic auto industry is also in great trouble. These have been serious blows to our house-and-car oriented economy. The current economic meltdown is not just an American problem, however; its impact is now being felt in the European Union (all of the Eurozone countries are in recession at this moment) and throughout the rest of the world.

As far as the U.S. is concerned, business as usual will no longer help us to get back on a path of economic growth once the current recession is over. We need to move beyond gas-guzzling SUVs, McMansions, and financial gymnastics if the U.S. is to regain prosperity, attain sustainable economic growth and remain competitive in today's highly competitive global economy.

Attaining a Culture of Innovation

There is wide agreement that innovation is the best, and perhaps the only way, for us to sustain our prosperity. Instead of simply creating new waves of the same old consumer products and services that continue to exacerbate our problems, we need to focus on transforming smart ideas that address real problems and needs into valuable products and services. *Innovation* needs to be our mantra from here on, practiced consistently and frequently, until it becomes part of our national psyche and culture.

Leading-edge science and technology have been America's strong suits since its inception. The country's culture of innovation has been a foundational factor, and the list of American innovation successes is long and impressive. Four of the eight winners of *The Economist* magazine's 2008 innovation awards are Americans. They include Jimmy Wales for Wikipedia, the free online encyclopedia; Steve Chen and Chad Hurley for YouTube, the popular video-sharing website; Arthur Rosenfeld for the promotion of energy efficiency; and Bill and Melinda Gates for developing a businesslike approach to philanthropy. Their foundation provides an enabling platform for non-profit organizations that are improving the lives of millions of people around the world.

For the U.S. to prosper in the 21st century, the nation must encourage the growth of world-class innovation regions. The good news is that we already have a number of them, including the Route 128 region around Boston, the Research Triangle in North Carolina, the high-tech region in Austin, and many others. But none surpass Silicon Valley in the San Francisco Bay Area of California in creating new businesses stemming from innovations, or in maintaining an impressive record of sustained progress for more than five decades. Silicon Valley undoubtedly is the "mother of all innovation regions" in the world.

Who is This Book For?

Understanding and replicating the Valley throughout the nation is of paramount national importance, particularly at this time of economic crisis. *Business leaders* across the country will benefit greatly from following the examples of successful companies in the Valley. *National and regional economic development leaders* would be wise to shape regions elsewhere along the lines of Silicon Valley, especially since the Valley's model now extends beyond high-tech industries. Understanding the key elements of what makes Silicon Valley tick is a must for *academic leaders*, as well as leaders of *think tanks* and *research organizations*, who play vital roles in the enhancement of innovation regions. It is also important for *venture capitalists and angel investors* to fully understand the vital roles they play, not only in providing the capital that is so vital for starting new businesses, but also in providing expertise and resources that are critical to the success of early-stage companies. Finally, it is vital for *government officials* at all levels to understand the critical role of policy, finance, law and resources in the emergence and enhancement of businesses in the innovation economy.

How This Book Came to Be

Continuous innovation has been at the heart of the Valley's prolonged success and resilience for nearly a century. My interest in understanding the Silicon Valley economy goes back several decades to the time when I worked for the Pacific Gas & Electric Company in San Francisco, one of the largest utilities in the United States. Silicon Valley is an important region for the company's energy sales. As PG&E's Chief Economist, I tracked the Valley's economy on a quarterly basis for more than a decade.

I have continued to track the Silicon Valley economy over the years since my PG&E days, first as a columnist for the Media News Group, and more recently as a Director for the worldwide consulting firm, LECG, LLC. Furthermore, in the process of writing *Dot-Com to Dot-Bomb: Understanding the Dot-Com Boom, Bust and the Resurgence* (2004), a book that analyzes the severe Boom/Bust cycle that the Valley experienced in 2000-2001, I came to understand how the Valley regained economic vitality after experiencing a severe economic upheaval.

This book builds on an earlier volume, *Silicon Valley: Ecology of Innovation*, which I was commissioned to write by the Science Parks Association of Spain (APTE) in 2008. A limited edition of that volume was published in English and Spanish and distributed directly by APTE to science parks officials and key stakeholders throughout Spain and abroad.

What Makes Silicon Valley Tick? is based on a series of papers and lectures that I presented at several conferences in Spain in late 2008 plus updated information on the state of Silicon Valley. My presentations in Spain included the APTE annual conference in Orense and the Red de Espacios Technológicos de Andalucia (RETA) annual conference in Malaga in November 2008. The general theme of these presentations was the critical role that innovation regions such as Silicon Valley play in 21st century economic development. The goal of these presentations was to provide science park executives with a framework for understanding Silicon Valley's economy, with the expectation that this would ultimately enhance the success of the scores of science parks currently operating in Spain.

This book answers many frequently asked questions put to me by representatives from many aspiring technology regions in the U.S. and in Spain over the last few years. They ask: What is Silicon Valley? How does the Valley's economy work? Why has this region been so successful? What underlies its resilience? What threats does it face, and what can be done about them? What can we learn from the Valley that we can apply to our region?

The Innovation Ecosystem

In seeking a framework for analyzing the economy of the greater Silicon Valley region, I concluded that the most effective framework was that of an ecosystem—an innovation ecosystem. Looking at an economy from a more organic, less mechanical perspective has been gaining popularity among economists. William Wulf, the researcher, entrepreneur and former president of the National Academy of Engineering, has used the phrase *ecology of innovation* to describe how various factors interact in the U.S. economy to enhance or hinder its ability to innovate. These factors, according to Wulf, include intellectual property law, tax codes, patent procedures, export controls, and immigration regulations.

In this book, we've taken a broader approach to identifying the key elements of the Silicon Valley's innovation ecosystem. Like every living ecosystem, Silicon Valley's very survival is the outcome of the complex interplay of a number of factors. Viewed as an innovation ecosystem, you can see Silicon Valley as a unique, lively creature which consumes and transforms knowledge and ideas into streams of innovative products and services through the continuous formation of new companies, within a complex matrix of relationships among various stakeholders in the region.

In developing our concept of the Valley's innovation ecosystem, we surveyed existing literature on Silicon Valley and applied our own experience, research and available data. We believe our framework provides a balanced and well-integrated model of the seven key elements that form the core of this highly successful innovation ecosystem. We offer this framework in hopes that it ultimately helps contribute to the prosperity of other regions, here and abroad.

Tapan Munroe

Silicon Valley: Fertile Ground for Innovation

Silicon Valley, in the southern San Francisco Bay Area, is the birthplace of many pioneering technology companies. It has sustained its place as the world's most dynamic innovation ecosystem for over a century.

Source: Silicon Valley Map

As of the last count there are nearly 59 technology regions that include the *Silicon* moniker.[1] Names range from Silicon Alley in New York City's lower Manhattan and Silicon Alps in Austria to Silicon Tundra in Ottawa, Canada and Silicon Wadi in Israel. This phenomenon reflects the enormous worldwide reputation and appeal of the Silicon Valley model of a successful technological region. The hope is always that the success associated with the *Silicon* moniker will somehow rub off on an aspiring region. Such is the strength of the Silicon Valley brand.

In my travels to other high-tech regions around the world, the questions I am asked most often are: What is Silicon Valley? How does the Valley's economy work? Why has this region been so successful? What underlies its resilience? What threats does it face, and what can be done about them? What can we learn from the Valley that we can apply to our region?

Ecosystem Basics

The key to understanding what makes Silicon Valley's innovation economy so successful can be understood via the workings of the unique evolutionary model: the biological ecosystem. Just like any biological ecosystem, a successful innovation ecosystem is based on key elements that adapt and evolve in the face of constant change. The ability of these key elements to rapidly adapt to change assures resilience in response to external shocks and internal upheavals. Resilience supports long-term sustainability so that the hard-earned prosperity of one cycle of success becomes the foundation for the next cycle, and so on. The success of an innovation ecosystem relies on the vitality and health of each of these elements and their interconnectedness. Conversely, a failing or unhealthy element can undermine the existence of the entire ecosystem.

Innovation

The process of commercializing an idea in the form of:

- A new or improved product
- A new production method
- A new business organization
- New uses for existing products
- New markets for existing products
- New distribution channels

Ecology of Innovation

The study of the key elements of an economy that foster and support innovation and their relationships to each other within that economy.

At the heart of this book are the key elements that make up the core of the Valley's innovation ecosystem. This thriving high-tech region in the southern San Francisco Bay Area is a unique economic environment where bold, new ideas compete and then are transformed into products and services that reshape our world. In this process of transformation, the Silicon Valley entrepreneur becomes part of a symbiotic web of supportive relationships. In a world of innovation, ideas need entrepreneurs to survive and entrepreneurs need ideas to flourish.

Modeling the Innovation Ecosystem

In Silicon Valley's innovation ecosystem, the following key elements, each playing different and mutually supporting roles, maintain and enhance the ecosystem's health:[2]

- World-class research universities
- Entrepreneurs
- Investment capital
- A talented, well-educated workforce
- Social and professional networks
- A favorable business climate
- A high quality of life quotient

They each contribute in many interconnected ways to the success of entrepreneurs' ventures, and in doing so, contribute to their own prosperity and the prosperity of the region.

Innovation and technological change are undoubtedly the main drivers of economic growth at organizational, industry, and macroeconomic levels. The fundamental point is that without innovation, there is no increase in productivity, and without productivity, there is no increase in prosperity.

For decades, Silicon Valley has sustained its success by consistently providing a steady stream of ideas and innovations that have created value for consumers and organizations worldwide. In this book we will help you understand what has made Silicon Valley so successful and what factors have contributed to its long-term sustainability as a world-class high-tech economy.

In order to do so, we must understand the Valley's ecosystem by identifying its key elements, in Part 1. In Part 2, we learn how they interact to sustain and enhance one of the most well-known and extraordinary regional economies in the world.

Finally, in Part 3, we examine the vulnerabilities of the Valley's ecosystem— after all, no ecosystem is immune to risks, threats and challenges to its continued survival and vitality.

Part 1

A Profile of Silicon Valley

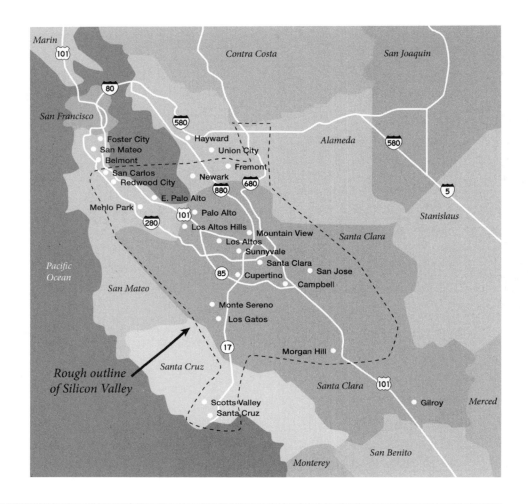

Rough outline
of Silicon Valley

San Francisco Bay Area Counties and Silicon Valley

Source: Joint Venture: Silicon Valley Network, Munroe Consulting, Inc.

A History of Innovation

F or more than a century, the southern region of the San Francisco Bay Area known as Silicon Valley has been a highly successful hub of invention and innovation.[3] Dubbed Silicon Valley by Ralph Vaerst, a Northern California entrepreneur, and popularized in 1971 by his friend Dan Hoefler, a reporter for the trade journal *Electronic News*,[4] the name has been widely embraced by the media and the public.

Santa Clara County and Silicon Valley

While there are no official borders of Silicon Valley, the term was originally used to define the area in and around the city of San Jose (often referred to as the Capital of Silicon Valley). Today, Silicon Valley includes Santa Clara County as well as adjacent parts of Alameda County, San Mateo County, and Santa Cruz County. (See map, opposite.)

Silicon Valley embraces 29 cities, including Palo Alto, San Jose, Mountain View, Sunnyvale and Fremont, though the region's economic base extends up to San Francisco. The region's economic influence extends for a full two-hour commute to the north into Marin, Napa and Sonoma counties, into the East Bay counties of Alameda and Contra Costa, and further east into San Joaquin County.[5] In 2006, 11 Valley cities ranked among the top 20 most innovative cities in the United States.[6] Figure 1 provides a snapshot of the region and its population.

Stanford University

The seeds of Silicon Valley's innovation economy were sown in 1891 when former California governor, U.S. Senator and wealthy railroad magnate Leland Stanford and his wife, Jane, opened Stanford University on the grounds of their

Figure 1

Profile of the Silicon Valley Region

Area:	1,500 square miles
Cities:	9
Largest City:	San Jose
Population:	2.44 million
Average Annual Wage:	$74,302
Education (Degree):	
Bachelor's	26%
Graduate/Professional	18%
Ethnicity:	
White, Non-Hispanic	42%
Asian	29%
Hispanic	24%
Black	3%
Other	2%

Source: U.S. Census Bureau, 2006

Palo Alto ranch.[7] Since then the region has grown to become one of the world's preeminent powerhouses for innovation. No other region has so consistently yielded such a prolific harvest of revolutionary technologies and world-class companies. (Figure 2) Many of these technologies have changed (sometimes radically) our way of living and, in the process, made many of the region's innovators, implementers and investors enormously wealthy. No other region has so successfully managed to repeatedly reinvent itself in the face of ever-changing technology, major economic upheavals, and rapid globalization.

While Stanford University played a catalytic role in the birth of Silicon Valley and it continues to be instrumental in supporting the Valley's success, Silicon Valley has, over the years, also been influenced by its proximity to other world-class research universities and think tanks. These include the University of California (UC) campuses of UC Berkeley, UC San Francisco, think tanks such as the Xerox's Palo Alto Research Center, and research laboratories such as Lawrence Berkeley National Laboratory, Lawrence Livermore National Laboratory and Sandia National Laboratory.

Figure 2

A History of Silicon Valley Innovation

1891	Stanford University founded
1909	Federal Telegraph Company founded
1912	Lee De Forest perfects vacuum tube
1938	Hewlett-Packard founded
1951	Stanford Industrial Park established
1956	Shockley Semiconductor Laboratory founded
1957	Fairchild Semiconductor founded
1968	Intel founded
1976	Apple Computer founded
1982	Sun Microsystems founded
1984	Cisco Systems founded
1994	Yahoo! founded
1995	Netscape IPO triggers the Dot-Com Boom; eBay founded
1998	Google founded
2000	Dot-Com Bust and Telecom Meltdown
2007	Rise of Web 2.0/3.0 and Cleantech

Source: Joint Venture: Silicon Valley Network

These highly-regarded universities and think tanks have consistently provided the Valley with a steady stream of ideas, inventions, engineers and entrepreneurs. In presenting the history of Silicon Valley, we are quite aware that our discussion of events, people and achievements is very Stanford-centric. We believe the Bay Area's University of California campuses and the UC system as a whole have played an equally important role in contributing to the development and success of Silicon Valley. We will focus on the UC system's role in a later chapter. In fact, Silicon Valley is much more of a San Francisco Bay Area regional economic entity than strictly a South Bay economy.

The popular story of Silicon Valley is to a great extent the story of the inventions, innovations and business ventures initiated by Stanford students, graduates, faculty, and friends. Stanford's open and entrepreneurial culture (influenced heavily by its founder) was a major factor supporting and encouraging the integration of business and education in a way that was considered heretical in academic circles. Even in this, Stanford University was a pioneer of innovation. That was because historically, the academic study of business was not considered an appropriate use of university resources and faculty time.

Sparks of Innovation

> Silicon Valley's heroes are the successful entrepreneurs who have taken aggressive professional and technical risks: the garage tinkerers who created successful companies.
>
> **AnnaLee Saxenian**
>
> *Source: The New Argonauts: Regional Advantage in a Global Economy*

It is common lore that the birth of what we recognize as Silicon Valley (both geographically and economically) began with the founding of the Hewlett-Packard Company in 1938. Others attribute the birth of the Valley to the formation of Stanford Industry Park in 1951, or to the relocation of the pioneering electronics firm Varian Associates from Santa Clara to the Park in 1953. Others credit the defection of a handful of engineers (the so-called "Traitorous Eight") from Shockley Transistor to form Fairchild Semiconductors in 1955.

In his book, *How the Silicon Valley Came to Be*, Timothy Sturgeon of Massachusetts Institute of Technology points out that these are certainly seminal events in the long history of the Valley, but it is important to recognize that the roots of today's Silicon Valley extend back to the turn of the twentieth century.[8] Hewlett-Packard and Varian were not the first important electronics companies to rise near Stanford, and Fairchild Semiconductor was not the region's first spin-off.[9]

Rather, the roots of today's worldwide electronics industry can be traced back to the pioneering innovations of two Silicon Valley technologists and entrepreneurs. The first was Stanford graduate Cyril Elwell, who developed a highly successful radio telephone system. Backed by David Star Jordan, the president of Stanford University, and C.D. Marx, head of the University's Civil Engineering Department, Elwell founded the Federal Telegraph Company (FTC) in 1909.[10] The second was Lee De Forest, a brilliant young innovator who perfected the vacuum tube in 1912 while working for FTC. De Forest's vacuum tube revolutionized the radio industry, literally launching the Age of Electronics.[11] In fact, as William Hewlett, co-founder of Hewlett Packard Company, has commented, "the beginning of Silicon Valley was a supernova [that] caused a rippling effect, setting the stage for future events."[12] Hewlett attributes this supernova to the young De Forest and his work at Federal Telegraph Company.[13]

The ripples from FTC's impact on the world of electronics resulted in the first wave of Silicon Valley spin-offs, including Magnavox, Litton Industries and Fisher Research Laboratories. Magnavox became a recognized leader in manufacturing a broad range of consumer electronics products and pioneered the first home video game system in 1972.[14] Harold Elliott, an FTC engineer, developed the core technology that was the foundation of RCA's groundbreaking "single dial tuner," the home radio set that hit the market in 1927.

That same year, a young Stanford instructor, Fred Terman, launched the University's first radio and electronics engineering program. Terman's enthusiasm for radio and electronics helped him befriend many of the budding industry's luminaries, including Cyril Elwell of FTC and Philo Farnsworth, who invented the television at his San Francisco laboratory. Under Terman's leadership, Stanford's Radio Engineering Laboratory (REL) became one of the nation's top academic electronics centers. From the very beginning, Stanford leaders envisioned the role of the University as helping the region to build a strong local economy. This sense of regionalism helped the University align its interests with those of local high-tech firms.[15]

Stanford's "Secret Weapon"

Out of this fertile intellectual ground came two Stanford graduates, William Hewlett and David Packard. Both were students in Fred Terman's electronics program. Terman's field trips to nearby electronics research labs and his stories about entrepreneurial inventors inspired Hewlett and Packard to launch the Hewlett-Packard Company (HP) in 1938 in a Palo Alto garage. Nearly 70 years later, HP is one of the world's largest information technology firms.[16] HP's early success, the end of the World War II and Stanford's growing reputation as a world-class university set the stage for the Valley's second supernova: the Stanford Industrial Park. In the late 1940s, Stanford was experiencing financial troubles stemming from the University's rapid growth following the end of World War II. Selling any part of the University's large land holdings was out of the question, forbidden by Leland Stanford's gift to the University. Instead, the University opted to lease its surplus land—a decision that Terman called "our secret weapon," and one that ultimately transformed the region's economy. It was Terman who suggested that leases be limited to high-tech companies with which there could be a mutually beneficial relationship with the University and its students.[17] This was a bold move that contributed significantly to the Valley's culture of business-academic cooperation. From a financial perspective, the University benefited greatly from royalties coming from products spawned from campus-supported research, and from interests held in spin-off companies resulting from commercial research conducted in campus labs.

In 1951, the Stanford Industrial Park (renamed in 1955 as Stanford Research Park) was opened with the goal of creating "a center of high technology close to a cooperative university." A number of iconic technology companies, including Varian Associates, General Electric, and Eastman Kodak, quickly signed leases. In 1952, IBM set up a key research facility nearby. Soon after, in 1956, Lockheed Corporation located an aircraft division in the Stanford Research Park. These cutting-edge companies, crafting their technological wonders in such close proximity, had enormous impact on the local economy. The heart of

Silicon Valley pulsed in Stanford Research Park, while many related and supporting companies flourished in surrounding communities.[18]

The Silicon Revolution

While Stanford Industrial Park was signing leases with its first tenants, William Shockley, a brilliant yet sometimes brash engineer, co-invented[19] the first commercially viable transistor in 1951, marking the beginning of the semiconductor revolution in electronics. At the urging of Fred Terman, Shockley moved to Mountain View in 1956 to head up the newly-founded Shockley Semiconductor Laboratory—a division of Beckman Instruments—which became Shockley Transistors Corporation in 1958.[20] Shockley was a genius, but he was also a difficult man to deal with, and while he was able to attract top talent because of his reputation, his management style made it difficult to keep his team together.

In one of the Valley's most historic events, Robert Noyce, Gordon Moore, and six other Shockley engineers[21] resigned en masse in 1957 to form Fairchild Semiconductor in Palo Alto. Known as the "Traitorous Eight" (or "the "Fairchild Eight" or "Shockley Eight"), these men left Shockley Semiconductor Laboratory in 1957 in a disagreement with Shockley over the use of silicon as the exclusive material for making semiconductors. Noyce, Moore and the others believed that designing a silicon-only semiconductor was an important technical opportunity to pursue, but Shockley disagreed and actively discouraged them from this line of research. Fairchild became the first company to mass-produce integrated circuits, thereby replacing Shockley's technology. Shockley sold his company in 1961 and joined Stanford University as a faculty member.

Fairchild Semiconductor also eventually became a parent to over forty local high-tech companies,[22] laying the foundation of another core value of the Valley's entrepreneurial culture: the belief that anyone with a good idea can find the capital, put together a team and go after a market opportunity.[23] By the 1960s, no other region had such a concentration of cutting-edge high-tech companies—

and no other region could boast of being the birthplace of the electronics revolution with so many firsts in the field. Yet the Valley's culture of innovation was brewing new and even more revolutionary inventions. In 1968, Noyce and Moore formed Intel (now the world's largest semiconductor company) in Santa Clara, roughly at the geographical center of Santa Clara County and the South Bay Area. In 1971, Intel created the first microprocessor, the platform technology for the coming microcomputer revolution.[24] In 1972, Eugene Kleiner co-founded Kleiner, Perkins, Caufield & Byers—one of the Valley's best-known venture capital firms. Sheldon Roberts, Jean Hoerni and Jay Last founded what became Teledyne, while Julius Blank co-founded Xicor. Victor Grinich became a professor at UC Berkeley and Stanford University. Other "Fairchildren" (as spin-offs of Fairchild Semiconductor are sometimes called) include National Semiconductor and Advanced Micro Devices (AMD).

Parallel to the evolution of Fairchild Semiconductor, the geniuses at the Xerox Palo Alto Research Center (PARC) were developing the first graphical user interface for computers, the platform technology for all of today's highly visual operating systems and software applications. No other region has so influenced the development of such fundamental technologies in electronics. But there's more. In 1973, Stanley Cohen of Stanford University and Herbert Boyer of UC San Francisco invented a technique for splicing genes, leading to the formation of the biotech industry.

The Digital Explosion

With all of the foundational hardware and software technology pieces at their disposal, two more young innovators, Steve Jobs and Steve Wosniak, put the pieces together to launch Apple Computer in 1976. In doing so, they totally changed the way we interact with computers (and now music and telephones). The Apple Macintosh graphical user interface was built upon work done by Doug Engelbart at Stanford Research Institute and by engineers at Xerox PARC. In 1982, Stanford graduate students Andy Bechtolsheim, Vinod Khosla and Scott McNealy plus Bill Joy founded Sun Microsystems as a spin-off of the

Stanford University Network (in fact, "Sun" is an acronym for "Stanford University Network").[25]

That same year, Jim Clark, an associate professor of electrical engineering at Stanford, founded Silicon Graphics, Inc. (SGI), with a group of seven Stanford graduates and research staff. SGI created the first graphic workstations.

In 1984, Leonard Bosack and Sandra Lerner, also from the Stanford University Network, founded Cisco Systems, the premier Internet router company in the world. With the creation of the World Wide Web in 1989 and the explosion of opportunities that this presented, Clark hired the Mosaic web browser pioneer Mark Andreessen in 1993 and formed Mosaic Communications, predecessor of Netscape Communications Corporation. Soon thereafter, Netscape's hugely successful initial public stock offering (IPO) in August of 1995 lit the fuse that launched the Dot-Com skyrocket. The Internet proved to be both a blessing and a curse for the Valley, as we will discuss later.

In 1994, Jerry Yang and David Filo, both PhD candidates in electrical engineering at Stanford, started a simple directory of websites that exploded into Yahoo!, one of the Web's most popular search portals. A year later, San Jose became home to eBay, the Web's most popular online auction site, founded by Pierre Omidyar, who earlier had worked for Claris, a software subsidiary of Apple Computer. The very next year, two more Stanford PhD students, Larry Page and Sergey Brin, formed Google, underwritten by a $100,000 check from Andy Bechtolsheim, co-founder of Sun Microsystems. Six months later, Google landed $20 million in equity financing from rival venture capital firms Kleiner Perkins Caufield & Byers and Sequoia Capital. Yahoo!, eBay and Google have become icons of the Internet Age, and their success has created many young multimillionaires as well as a few billionaires.

Innovation and Evolution

One can easily see in this brief historical overview that Silicon Valley's success was not the result of a well-planned economic development policy. Nor did the Valley's high-tech industry spring full-blown from a single event. Its success has evolved over 100 years, a result of the synergy among the region's well-established entrepreneurial culture, the influences of a handful of visionary and entrepreneurial leaders, and the pioneering innovations of many talented individuals who leveraged their ties with a world-class university. In this way, they changed the world with their revolutionary technologies, and made their fortunes while doing so.

Stanford University's entrepreneurial approach to integrating teaching and research with business once ruffled feathers in more traditional academic circles.[26] Eventually, though, it proved invaluable as an inspiration for generations of entrepreneurs and engineers who, to this day, maintain a culture which fosters the open exchange of ideas, friendly collaboration and world-class creativity. The sheer number of cutting-edge technology companies that were drawn to and spawned from the Stanford Research Park is a testament to Fredrick Terman's vision of a "center of high technology close to a cooperative university."

Sixty years after the opening of Stanford Research Park, the Valley is still rich with entrepreneurial spirit and innovation, even though the region's economy has undergone massive change and experienced much hardship as well as success over that time.

Our next look at Silicon Valley will be from a more traditional perspective.

The Silicon Valley Economy

I n the past 25 or so years, Silicon Valley's world-class innovation economy has experienced dramatic ups and downs. This is not surprising: a highly innovative economy is risky business. The Valley economy has soared as new technologies replaced old ones, then slowed as once cutting-edge products became commodities —until the "next big thing" arrived on the scene.

A World-Class Innovation Economy

The most dramatic economic rise and fall occurred in the Boom/Bust cycle of 1998–2001, when the region's economy rode the Dot-Com rollercoaster.[27] Despite the volatility in the Valley's economy, the region has enjoyed average annual real economic growth of 5% since 1978—a full 1% greater than in California, and 2% greater than the U.S. growth rate for the period. With the Dot-Com Bust of 2000, personal income declined in Santa Clara County and in the Bay Area for two straight years—a decline of more than $7 billion to the region's economy. Growth returned to "normal" again by 2004 and out-paced state and national growth at 5.7% to 6.4% through 2006.

The Valley's economy recovered in the period 2004-2007, but once again it is being tested by the financial and economic meltdown of 2008-2009. The region is resilient but not immune to severe economic downturns. Nonetheless, it is expected that the region will perform relatively well despite the worldwide recession. We discuss the Valley's ability to survive economic shocks in Part 3.

Silicon Valley's San Jose area currently leads other high-tech regions of the world in terms of "knowledge competitiveness." (Figure 3) Based on an index established by the Center for International Competitiveness, which is made up of nineteen "knowledge economy" indicators that include per-capita jobs in

Figure 3

World Knowledge Competitiveness Index
Top Ten Regions

1	San Jose, California, U.S.
2	Boston, Massachusetts, U.S.
3	San Francisco, California, U.S.
4	Hartford, Connecticut, U.S.
5	Seattle, Washington, U.S.
6	Grand Rapids, Michigan, U.S.
7	San Diego, California, U.S.
8	Stockholm, Sweden
9	Rochester, New York, U.S.
10	Los Angeles, California, U.S.

Source: Center for International Competitiveness, 2005

intellectual property development, per-capita patent registrations, and per-capita venture capital dollars invested, San Jose was rated the top "knowledge economy" worldwide, followed by Boston, San Francisco, Hartford, and Seattle. In terms of per-capita jobs in intellectual property development and per-capita venture capital dollars invested, Silicon Valley also ranked first in the world over prominent Asian rivals such as Tokyo, Shanghai, Beijing, Seoul, Singapore, Taiwan, and Bangalore.[28]

Industry Clusters and Specialization

One of the aspects of Silicon Valley's economy that is indicative of a mature innovation economy is the strength of its high-tech industry clusters. Michael Porter of Harvard University popularized the notion of industry clusters in 1990.[29]

Industry clusters are geographic concentrations of competing and collaborating companies in the same or similar fields of business, as well as the specialized suppliers, service providers and other related businesses and institutions that support these core companies. A cluster emerges when a "critical mass" of core industry companies locates in an area in order to capitalize on local competitive advantages. A cluster becomes established when its aggregate employment reaches a significant level and, as a group, the core businesses become a major contributor to the local economy.

The 19th-century economist Alfred Marshall[30] suggested that when businesses in the same or complementary industries locate in the same region, they improve their access to resources of all kinds (e.g., skilled labor, business and technical expertise, etc.), thereby enhancing their competitiveness. Porter argues that competition is the driving force behind cluster development. As a competitive business grows, it generates demand for related industries. Competition between rival firms in the cluster forces them to be innovative. This spurs new technologies, spin-off businesses, more research and development, and new skills and services.

Businesses within an industry cluster generally draw from the same pool of skilled workers, who often move from one firm to another within the cluster. This free-flow of skilled workers becomes a channel for transferring knowledge and skills, further promoting competition and growth. That growth leads to what Porter calls "horizontal clustering," as new technologies and skilled labor move to related industries in different sectors of the regional economy.

A measure called *employment concentration* compares the percentage of employment in a regional cluster to the percentage of employment in the same cluster nationally. An employment concentration that is greater than the national percentage for a regional cluster indicates that it is what is called a traded industry. This means that this cluster is reliant on markets outside the local economy for maintaining its competitive advantage. The so-called traded industries account for nearly 30% of regional jobs in the U.S. In addition they have higher wages, higher productivity, and higher rates of innovation.

Looking at employment concentrations, data suggests that the Valley's economy includes seven high-tech industry clusters: semiconductors, software, computer and communication hardware, innovation services, biomedical, electronic components, and creative services.

Figure 4

Silicon Valley Cluster Employment Concentration Relative to the U.S.

Values greater than 1 indicate concentrations greater than the U.S. as a whole

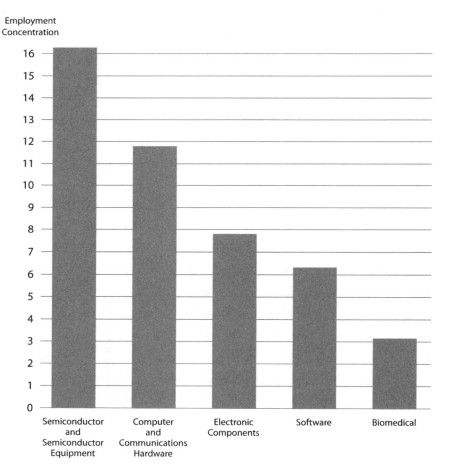

Source: Economy.com

Of these, five have higher employment concentrations than other major high-tech regions in the U.S.: semiconductors, computer and communications hardware, electronic components, software, and biomedical. (Figure 4) By definition, these five clusters are traded industries and have higher wages, productivity, and innovation levels than the U.S. economy-wide average for the same industry sectors. This explains the broad level of technological competency of the Valley relative to other regions such as Austin, Boston, and Raleigh-Durham.[31]

Many studies have analyzed and compared the country's top regional economies in order to determine why and how these regions developed and maintain their individual levels of prosperity. In 2001, the Brookings Institute looked at fourteen of the nation's top regional economies.[32] One of their areas of study examined "principal product specializations." Their research showed that while most of these top economic regions excelled in two product specializations, only the Boston, Washington, D.C. and Austin regions showed specializations in three principal product fields. Meanwhile, Silicon Valley excelled in seven: semiconductors, computers, software, communication equipment, semiconductor manufacturing equipment, electronic design automation software, and data storage. This broad-based excellence of Silicon Valley has been a key factor in the region's global competitive advantage as well as its long-term viability. (Figure 5)

Economic Performance

Looking back on 1977-2007 figures, the Silicon Valley economy has grown at an average annual rate of 5%—a full 1% greater than California's average rate of 4% and 2% greater than the U.S. rate of 3%.[33] (Figure 6) But this "average rate" masks a story of much more dramatic fluctuations in economic fortune. Coming out of the 1970s, the Valley was flying high with an economic growth rate of more than 10%. While the nation stumbled on the oil crisis of the early 1980s and fell into a recession, the Valley's economy kept on humming along, though at a steadily slowing pace. At that time, oil was not the driving factor in the Valley's economy as it was to the nation as a whole. Semiconductors were, though, and the Valley had other cost issues to contend with.

Figure 5

U.S. Regional Product Specializations

Atlanta	Databases, (Telecommunications)
Austin	Semiconductors, Computers, SME*
Boston	Computers, Medical Devices, Software, (Biotechnology)
Denver	Data Storage, Telecommunications Equipment and Software
Minneapolis–St. Paul	Computers, Peripherals, Medical Devices
Phoenix	Semiconductors, (Aerospace)
Portland	Semiconductors, Display Technology, SME, EDA,** Silicon Wafers
Raleigh-Durham	Computers, Databases, (Pharmaceuticals)
Sacramento	Computers, Semiconductors
Salt Lake City	Software, Medical Devices
San Diego	Communications Equipment, (Biotech)
San Jose	Semiconductors, Computers, Software, Communication Equipment, SME, EDA, Data Storage
Seattle	Software, (Biotechnology, Aerospace)
Washington D.C.	Databases, Internet Service, (Telecommunications, Biotechnology)

Note: Parentheses indicate secondary specializations
*SME = semiconductor manufacturing equipment   **EDA = electronic design automation software
Source: Brookings Institute, 2001

Coming out of the 1981-82 recession, the Valley's growth was still well over 5%, but the cost of manufacturing semiconductors was soaring. The Valley's semiconductor industry was struggling to find a new business model. Economic growth came almost to a halt in 1986 (with only 0.5% growth) before innovative chip designers found a way to partner with Asian chipmakers in a symbiotic relationship. Out of the ashes, the *fabless* semiconductor business model was born. A fabless company (the word is compressed from fabrication-less) only

Figure 6

Economic Growth

U.S., California and Santa Clara County, 1978-2007

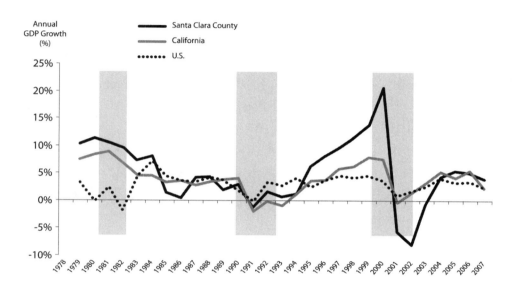

Note: Gray panels indicate U.S. economic recessions and slowdowns. Data for 2007 is forecast.

Source: Bureau of Economic Analysis, Munroe Consulting, Inc.

designs, develops and markets its chips, outsourcing the manufacturing stage. Valley firms proved that by outsourcing chip manufacturing to Japan, Taiwan and elsewhere, they could protect profits.

But just as the Valley's moguls thought they had dodged a bullet and economic growth had returned to a respectable 4.1% in 1987, the Cold War ended. By 1990, many Valley businesses were faced with the loss of key defense contracts—the lifeblood of many electronics and high-tech firms like Lockheed

Corporation (now Lockheed Martin). National defense spending declined by about 16.9% between the final Reagan Administration defense budget (FY 1989) and the final George H.W. Bush Administration budget (FY 1993).[34]

Subsequently, defense spending declined by another 18% between FY 1992 and FY 1997 under the Clinton Administration.[35] Given California's heavy involvement in defense-related industries in the Bay Area, as well as in the Los Angeles and San Diego areas, California's economy experienced three years of decline between 1991 and 1993 (-1.9%, -0.1% and -1.0% respectively).[36]

It wasn't the U.S. Cavalry that saved the Valley, however… it was the U.S. consumer. By the mid-1990s, the consumer electronics revolution had taken hold and sales of high-tech gadgets, particularly personal computers (PCs), soared. Annual PC sales grew by 20% a year in the early 1990s. Sales hit 18 million units in 1994, up 23% over 1993.[37] At the time, the market research firm IDC predicted that sales would continue to grow at a 15% to 20% clip in 1995.

But 1995 would turn out to be a good year in Silicon Valley—a very good year indeed. Setting the stage for one of the most dramatic economic booms in history were two key events. The European Organization for Nuclear Research, abbreviated as CERN, announced that its previously internal communication network called the World Wide Web would be open and free to anyone.[38] And Mosaic Communications Corporation announced the release of the Netscape Navigator Web browser in November 1994. The Age of the Internet was born. But the pivotal point in time was in August 1995, when Netscape's IPO stunned both the financial markets and the tech community. The Dot-Com Boom took off like a skyrocket… and so did the Valley's economy.

Supporting and even driving the surge in personal computing and Web browsing was the release of Microsoft's Windows 95 (Win95) in August of 1995—the first operating system to provide PC users with an experience similar to Apple's popular Macintosh. While the first releases of Win95 did not include an Internet browser, subsequent releases did. Microsoft had spoken, and Web browsing became a new national pastime. The post-Cold War doldrums of the early 1990s became a faded memory as the magic of the personal computer and all of its cousins (e.g., the PDA, the inexpensive cell phone) captivated the world. From 1994 to 2000, the Valley's economic growth rate soared from 1.2% to 20.8%.

It was like a dream, a boom that would never end. But it did. Just as August 1995 was a pivotal time in the Valley's history, so was March 2000, when the Dot-Com bubble burst and over a trillion dollars in equity vaporized almost overnight. The Valley's economy dropped precipitously from its peak in 2000 by 5.9% in 2001, another 8.1% in 2002 and then another 0.4% in 2003. Such turmoil in one of the nation's most powerful economic engines had to have broad-reaching impact. And it did: Both the state and the country saw a decline in economic growth, though not as significantly as the Valley's. And both the California and national economies recovered more quickly, as the Valley's entrepreneurs and investors retreated to their respective corners to lick their wounds, count their remaining pennies and figure out what to do next. It wasn't until 2004 that the Valley's economy would grow again, this time inspired by visions of new fortunes from clean energy, green technology, the highly interactive Web 2.0, nanotechnology, and new advances in biotechnology.

Employment

When we look at a profile of Santa Clara County from the perspective of employment, we see a markedly different pattern than is typical of the state and the nation. (Figure 7)

Of course, bundled together, the services subsectors—including professional and business services, education and health services, and leisure and hospitality, represent the largest single primary employment in the county, as it does statewide and nationally. But when we look at a more representative profile of the county by breaking out services into more discrete subsectors, we can clearly see how different the region's economy really is.

Despite the steady trend toward outsourcing manufacturing labor overseas, manufacturing generates the largest share of non-farm jobs in Santa Clara County (19.2% in 2007). In contrast, manufacturing jobs make up only 10% of

Figure 7

Employment by Sector Shares
U.S., California and Santa Clara County, Forecast for 2007

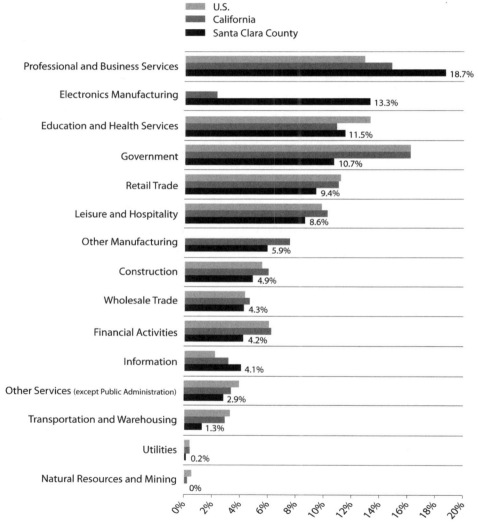

Legend:
- U.S.
- California
- Santa Clara County

Sector	Santa Clara County
Professional and Business Services	18.7%
Electronics Manufacturing	13.3%
Education and Health Services	11.5%
Government	10.7%
Retail Trade	9.4%
Leisure and Hospitality	8.6%
Other Manufacturing	5.9%
Construction	4.9%
Wholesale Trade	4.3%
Financial Activities	4.2%
Information	4.1%
Other Services (except Public Administration)	2.9%
Transportation and Warehousing	1.3%
Utilities	0.2%
Natural Resources and Mining	0%

Note: Data for the U.S. is not available for the Electronics Manufacturing or Other Manufacturing subsectors.

Source: Bureau of Labor Statistics, Munroe Consulting, Inc.

state and national employment. A closer look shows that electronics manufacturing represents a lion's share of the Valley's manufacturing jobs (69.1% in 2007), dwarfing the state average of 2.3%.[39] In fact, with a 13.3% share of the county's employment, the electronics manufacturing ranks second only to professional and business services as the region's top jobs sector.

One would think that the information sector would be a major source of jobs in Santa Clara County. Certainly it is a high-profile sector, as the county is home to eBay, Adobe and other internationally prominent software companies. But the professional and business services sector provides a much larger share of the county's jobs (18.7% in 2007). This, too, is a significantly larger share than we see across the state (14.9%) or around the country (12.9%). The information sector provides only 4.1% of the county's jobs, but this too is above the state and national averages of 3.1% and 2.2% respectively.

It is important to note that government provides an unusually smaller number of jobs in major entrepreneurial regions, and Silicon Valley is no exception. Federal, state and local governments provide only 10.7% of Santa Clara County's jobs (2007)—much lower than what we see at the state and national levels (16.2% for both). Clearly, high-tech manufacturing has been a key sector of the Valley economy. The sector has weathered numerous economic shifts and shocks, yet has maintained its position as a major source of employment and wealth-generation. Employment in electronics manufacturing peaked in 1986 at 186,000 jobs, then fluctuated over the next fifteen years until peaking again in 2000 at 179,800. But looking at the electronics manufacturing sector over the last seven years, we see a picture of steadily declining employment, with little evidence that jobs in this sector will ever recover to levels seen before the Dot-Com Bust.

In most regions of the U.S. economy, the share of manufacturing jobs as a total of all employment has been declining for decades, and the story is similar for the Valley. More than 64,000 jobs have been lost in electronics manufacturing since 2000 and projections call for steadily declining employment through 2020.[40]

Figure 8

Santa Clara Employment
Selected Sectors, 1978-2007

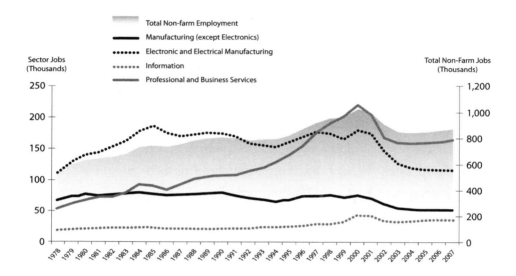

Note: Santa Clara County is considered as a surrogate for Silicon Valley in this analysis.
Source: Bureau of Labor Statistics, Munroe Consulting, Inc.

Jobs in non-electronics manufacturing have also been declining steadily since the late 1980s, but this trend is expected to level off over the next decade.[41] What we are seeing here is the Valley's steady shift from a manufacturing-based economy to a knowledge-based economy, similar to what we are seeing in most high-tech regions in the U.S. This is a result of a combination of factors including rising productivity in manufacturing, the elimination of many jobs through automation, and the migration of manufacturing jobs to lower labor cost countries such as China, India, Taiwan, and several eastern European countries.

Evidence of this structural shift is the steady growth in the professional and business services sector over the last thirty years. (Figure 8) Employment in this sector has more than tripled since 1977. Certainly, this sector rode the Boom/Bust rollercoaster up from 140,000 jobs to a peak of 220,000 in 2000, then down to 158,000 in 2004. But as of 2006, employment had returned to pre-boom levels (163,000) and forecasts suggest steady gains averaging 2.5% per year into the future.[42]

Jobs in the information sector also doubled in the last thirty years. Growth in this sector has averaged 2.7% per year, despite the Boom/Bust surge and decline. Employment in the sector has returned to mid-1999 levels and forecasts suggest modest but steady growth (1.3%) going forward.

Part 2

Silicon Valley's Innovation Ecosystem

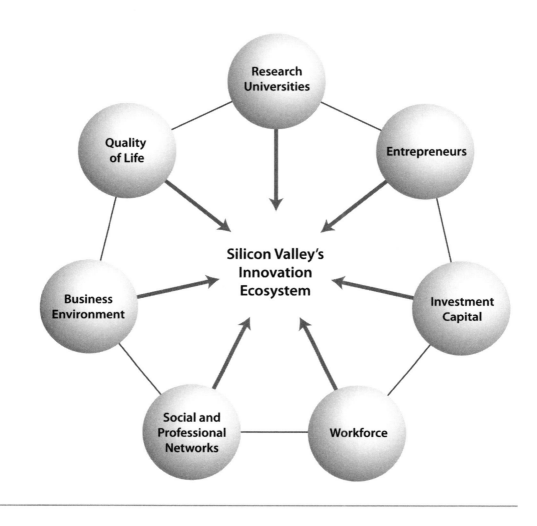

The Seven Key Elements of Silicon Valley's Innovation Ecosystem

A s we noted in the Introduction, the success of Silicon Valley springs neither from the brain of a brilliant planner nor the grandiose vision of a forward-thinking economic development agency. The evolution of the Valley as a global center for high-tech innovation was almost totally organic, rather than intentional. Certainly, the creation of the Stanford Industrial Park was an intentional effort to provide a center of high technology located near a university that shared its vision, but virtually every other aspect of the emergence of the Valley's prominence was unplanned, unexpected and unintended on the macro level. Its evolution represents a long series of individual initiatives, great innovations, and interactions among the stakeholders via ever-expanding social and professional networks.

James F. Moore formally introduced the concept of a business ecosystem in a *Harvard Business Review* article in 1993.[43] He described a business ecosystem as:

> [An] economic community supported by a foundation of interacting organizations and individuals—the organisms of the business world. This economic community produces goods and services of value to customers, who are themselves members of the ecosystem. The member organizations also include suppliers, lead producers, competitors, and other stakeholders. Over time, they co-evolve their capabilities and roles, and tend to align themselves with the directions set by one or more central companies. Those companies holding leadership roles may change over time, but the function of an ecosystem leader is valued by the community because it enables members to move toward shared visions to align their investments and to find mutually supportive roles.

Major companies such as Hewlett Packard, IBM, SAP, Microsoft, Softbank, and Intel initially used this economic community concept for business strategy development. Lately it has been more broadly applied to many problems, including foreign policy as well as economic development strategy development.

Ecology: The study of the relationship between organisms and their environment.

Ecosystem: A complex set of relationships among living resources, habitats, and residents of an area. When an ecosystem is healthy and is in balance it is sustainable. Diversity is a contributing factor to the health of an ecosystem.

Key Ecosystem Elements

So what are the fundamental elements that have served and sustained Silicon Valley's economy so successfully for so long? We have identified seven key elements[44] of the Valley's innovation ecosystem:

1. Research Universities

World-class research universities such as the University of California (UC) at Berkeley, UC San Francisco, and Stanford University form the foundation for a world-class innovation economy in many ways: via the generation and licensing of intellectual property; through the involvement of faculty as consultants and advisers to businesses; by supporting faculty in capitalizing on their innovations; by providing the private sector with a steady supply of talented engineers, designers, managers, etc.; by providing innovators with access to cutting-edge laboratories and equipment; and by encouraging a continuous dialogue among industry experts, faculty and students.

2. Entrepreneurs

In the innovation ecosystem, the entrepreneur is the biological host. Without the unique talents, traits and tenacity of the entrepreneur, bold new ideas would never see the light of day. We all have lots of new ideas, but the entrepreneur, driven by the energy and excitement of the core idea that is the seed of innovation, as well as by a hefty dose of self-interest and visions of personal gain, makes the commitment and takes the risk to manifest the innovation as a new product or service. A culture of entrepreneurialism and a tradition of serial entrepreneurship are key features of the Valley's ecosystem.

3. Investment Capital

Very few high-tech ventures can launch and grow to become world-class companies without large infusions of cash at crucial stages of development. While the Valley offers relatively convenient access to high net-worth individuals (sometimes called angel investors), multimillion dollar venture capital firms, and top investment banks, the brutal competition for cash is one of the primary survival tests for both innovations and entrepreneurs. (Theoretically, this competitive process should assure that only the best innovations make it to market, but as we have seen, this has not always proven to be true.) Investors do more than just provide money; they also offer a wealth of technical expertise, business experience and valuable connections to resources and people.

4. Workforce

No business venture can thrive without a skilled and dedicated workforce. The Valley is a magnet for talent from all over the world, and the diversity of the region's workforce has been a source of its strength and success. More so even than the social environment, the Valley's business environment is a melting pot of people and ideas from a wide range of ethnic backgrounds, academic disciplines, business cultures, etc. At the same time, as in every ecosystem, organisms (i.e., skilled workers) at every level are opportunistic. In a culture where job-hopping is an accepted practice, retaining talented employees is challenging for every company. As a result, Valley companies are experimenting with various innovations in the workplace to keep their employees loyal and happy.

5. Social and Professional Networks

Just as money is the primary fuel for innovative ventures, information is vital for survival and success, particularly when competing on a global scale. Information takes many forms and comes from a myriad of sources, including both formal and informal social and professional networks. Information, ideas, contacts and connections flow freely despite the hyper-competitive spirit that pervades the Valley. The degree to which the Valley's business community has successfully managed the tension between collegiality and competitiveness has proven to be a major contributor to the unique culture of its business environment.

6. Business Environment

Every ecosystem is dependent on the surrounding environment. That environment may be nurturing and supportive of vitality and growth, or it can be a source of stress and hinder or even threaten the long-term viability of the organisms within it. The same is true in an economic ecosystem. A region's economic environment includes many complex and interdependent factors: its social framework and political structure, its physical and economic infrastructures, its population profile, etc. In these ways and more, the Valley has created a healthy business environment for innovation.

7. Quality of Life

Business is business, but we're all human, and it's safe to say that world-class innovators, investors, and workers all appreciate a world-class lifestyle. The overall quality of life in Silicon Valley and the Bay Area is a significant contributor to the region's long-term economic success. With its comfortable Mediterranean climate, scenic beauty, first-rate cultural venues, cosmopolitan ambiance and proximity to sun, surf and snow, the San Francisco Bay Area easily qualifies as one of the most attractive regions in the United States. While quality of life may not seem the most important element of an innovation ecosystem, we believe that the Valley's locale has played a key role in its birth, evolution and long-term success. The people of California and the Bay Area have worked hard to protect their natural treasures and encourage a wide range of artistic and cultural expression, and the region as a whole benefits from it.

Innovation in an Interactive Web

As in any biological ecosystem, these key elements work in concert—no one element stands on its own. And, just like in nature, these elements form a web of relationships, interacting with each other in synergistic ways that strengthen the overall economic environment and contribute to the Valley's resilience, sustainability and long-term survival as a hub of innovation. While it may seem that one or two elements—research universities and venture capital, for example —are dominant factors in maintaining the balance of the system, seemingly lesser elements such as quality of life and business environment are also highly influential. Venture capitalists choose where they live and work and appreciate a region with cultural and recreational assets—they have neither the need nor desire to compromise their tastes. A research university may be top in training engineers, but if a region's business environment or culture does not encourage entrepreneurship (think of pre-1990s China), talented innovators will relocate to more supportive areas like the U.S.

As we will see, some or all of these same elements may be found in other regions, but the unique qualities and richness of these elements in Silicon Valley are the sources of its sustained economic vitality. For example, other regions of the country have an abundance of skilled workers (e.g., Detroit), but for various reasons, the industries that employ them like the auto industry and the surrounding social environments have not created cultures of life-long learning nor encouraged the kind of adaptability that is innate in Silicon Valley's workforce. Other regions (e.g., Spain) may have an abundance of entrepreneurs, but they may approach starting a business as a form of lifetime employment rather than taking the more aggressive start-grow-sell path of the serial entrepreneur. As we will see, serial entrepreneurs are catalysts and active agents in the social and professional networks that inspire and support new entrepreneurs, thus fueling the innovation economy. Without them, a region's networks lack the vitality to stimulate and support world-class business leaders. In the next section, we will take a closer look at each of these elements individually in the context of the Silicon Valley's economy, then examine the Valley's economy as a whole to see how it has survived, adapted and thrived for so long as the world's most dynamic economic ecosystem.

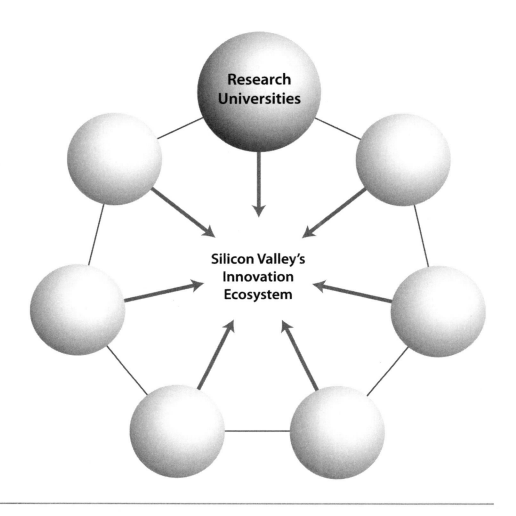

Key Element 1: Research Universities

An innovation ecosystem needs at least one if not several primary sources of intellectual property (IP), the essence of innovation. Research universities, think tanks, and corporate labs are important incubators for new ideas and innovation. They also provide critical training for the region's workforce and offer companies access to high-tech labs and equipment. Technology transfer programs help facilitate the commercialization of IP and bring returns in the form of royalties.

Research Universities

Research universities play a pivotal role in the evolution and success of innovation economies. The Silicon Valley economy has continued to flourish partly because the region is home to several world-class research universities. Earlier we discussed the pioneering role that Stanford University played in the evolution of the Valley. We would be remiss if we did not discuss the vital role that the University of California (UC) system has played in the development of the Valley, particularly in the last three decades. Thus we will focus on the contribution of the University of California system in this chapter.

The Impact of the University of California System

As the Valley's global reach and influence grow, it is important to keep in mind that Silicon Valley continues to transcend the geographic boundaries outlined at the beginning of this book. The Valley's economy has become borderless as a result of advances in communication and formal and informal linkages with other California regions, particularly with the various members of the ten-campus UC system. Nonetheless, we place special emphasis in this discussion on the contributions made by UC Berkeley. In many ways, UC Berkeley has been as influential on the evolution of Silicon Valley's innovation ecosystem as Stanford University, though it is less known and publicized.

The UC system has for decades contributed considerably to the high-tech economies of the three major urban regions of California: the San Francisco Bay Area; the Los Angeles Basin; and San Diego. UC has helped the growth of Silicon Valley and the state in two principal ways: supporting the key industry clusters in the region by training the region's workforce; and transferring knowledge, innovations, and new technologies from its campuses to private-sector start-ups and established companies.

Consider these two areas of contribution:

1. Support for Key Industry Clusters

The five urban-based industry clusters critical to the success of Silicon Valley include biosciences (pharmaceutical firms, medical laboratory research, and biomedical instruments manufacturing); computers and semiconductors; information technology; telecommunications (communication services and equipment manufacturing); and aerospace. UC provides one of the broadest ranges of education of any university system in the world—more than 150 fields. Its academic programs have been rated in the top 10 among U.S. universities for many years. The contribution of the UC-trained workforce to the region's industry clusters has been significant. Between 2002 and 2011, the UC system will graduate more than 34,000 specialists in science and technology. The total economic contribution of UC graduates to the region's clusters between 2002 and 2011 is projected to exceed $7.4 billion.[45]

> In the natural ecosystem of a tree, the branches and leaves won't get any water without a well-developed root system. In the Innovation Ecosystem, without a thriving research community at the roots, long term, sustainable advancement is not possible.
>
> **Judith Estrin**
>
> *Source:* Closing the Innovation Gap: Reigniting the Spark of Creativity in a Global Economy

2. Transfer of Knowledge, Innovations and Technologies

Intellectual property (IP) is the lifeblood of innovation. The region's easy access to IP—a major advantage for Silicon Valley's economy—is a direct result of the Valley's proximity to major research universities. Through aggressive technology transfer programs, the region's universities have provided many fledgling firms and established companies with access to foundational technologies.

The University of California has been the leader among U.S. universities in the creation of new patents (a key measure of intellectual property generation) for the last fifteen years. Figure 9 lists the number of patents received by U.S. universities in 2005. Topping the list is the University of California with 390 patents; followed by MIT with 136; then the California Institute of

Figure 9

Top Patenting U.S. Universities

U.S. University	Rank	No. of Patents
University of California	1	390
Massachusetts Institute of Technology	2	136
California Institute of Technology	3	101
Stanford University*	4*	90
University of Texas*	*	90
University of Wisconsin	5	77
John Hopkins University*	6*	71
University of Michigan*	*	71
University of Florida	7	64
Columbia University	8	57
Georgia Institute of Technology*	9*	43
University of Pennsylvania*	*	43
Cornell University	10	41

* Indicates a tie in the ranking among two or more U.S. universities.

Source: U.S. Patent and Trademark Office, 2005

Technology with 101; and Stanford University with 90. Patented processes and technologies, developed at UC campuses in San Francisco, Berkeley, San Diego and Los Angeles, and transferred to the private sector via aggressive and well-managed licensing programs, have had significant impact on the depth and breadth of the innovative products and services produced by Silicon Valley companies. This is particularly true in the telecommunications, computer technology, and biotechnology sectors.

One of the most effective channels by which UC-generated innovations find their way into the Silicon Valley, Bay Area, Los Angeles and San Diego economies has been through new business start-ups. For the purposes of illustration, let us consider electronics manufacturing start-ups in California. Electronics manufacturing is a key high-tech industry in the state, and the success of electronics manufacturing firms is critical to the success of related high-tech sectors. Innovations and advances in semiconductor technology often result in innovations and advances in downstream industries such as computers, scientific instruments, and communications equipment.

A 2007 study concluded that 64 of 424 (15%) of the electronics manufacturing firms in California were founded by graduates, post-docs or faculty of the UC system.[46] (Figure 10) This data is most likely understated since the academic affiliations of only 185 firms were available to the research team. It is noteworthy that UC engineers and scientists started seven of the twelve largest semiconductor firms in California. The list includes high-tech icons such as Intel, Advanced Micro Devices, Qualcomm, Broadcom, SanDisk, Maxim Integrated Products, and Marvel Technology Group. In 2006, these firms were among the top 50 semiconductor firms in the world.[47] In addition to spawning some of the world's major electronics manufacturing firms, UC scientists, engineers and graduates have formed a significant number of innovative start-ups in the industry over the last several years. Results of a study for the period 2000-2005 suggest that UC scientist or entrepreneurs in the semiconductor manufacturing industry started 42 new firms. Twenty-eight out of those firms had ties to UC Berkeley and, not surprisingly, were located in the San Francisco Bay Area.[48]

Figure 10

Leading California-Based Electronics Manufacturing Firms Founded by UC Graduates, Scientists, and Engineers

Company	Industry	Founder(s)	UC Affiliation	UC Campus
Advanced Micro Devices	Semiconductors	Jack Gifford	Graduate	UCLA
Affymetrix	DNA Chips	Stephen Fodor	Postdoc	UC Berkeley
Atheros Communications	Semiconductors	Teresa Meng	Graduate	UC Berkeley
Broadcom	Semiconductors	Henry Samueli, Henry Nicholas	Graduates, Faculty	UCLA
Cadence Design Systems	Electronic Design Automation	Richard Newton, Alberto Sangiovanni-Vincentelli, James Solomon	Graduates, Faculty	UC Berkeley
Intel	Semiconductors	Gordon Moore	Graduate	UC Berkeley
Qualcomm	Semiconductors	Irwin Jacobs, Andrew Viterbi	Faculty	UC San Diego and UCLA
SanDisk	Semiconductors	Sanjay Mehrotra	Graduate	UC Berkeley
Synopsys	Electronic Design Automation	Richard Newton	Graduate, Faculty	UC Berkeley

Source: UC Industry-University Cooperative Research Program

The 63 founding members of these firms represent significant ethnic diversity as suggested by their names—34 (nearly 54%) were of Asian descent, mostly of Chinese and Indian backgrounds. This underscores the tremendous success of Asian scientist and engineer entrepreneurs in the Silicon Valley and Bay Area. Further, the data once again illustrates the importance and intrinsic value of attracting the best and brightest talents, regardless of their ethnicity or national origin.

The data in the UC study suggest other important conclusions relating the UC system to the success of high-tech economies of Silicon Valley and the Bay Area:

- UC Berkeley scientist-entrepreneurs established more electronics-manufacturing firms in California than those originated by graduates from Stanford, MIT, and the Indian Institute of Technologies (located in more than a dozen cities in India) combined.
- The Bay Area and the Los Angeles and San Diego regions are the main centers for entrepreneurship in semiconductors and related industries in the United States. As a result, these regions receive nearly 70% of all the venture capital invested in semiconductors and related industries in the United States.
- The solid-state research and teaching programs at UC have been the major strength of the semiconductor and electronics manufacturing industry in the state. These programs started out in the late 1950s and 1960s at UC Berkeley and UCLA with a focus primarily on silicon-based technologies. In the second half of the 1970s, UC San Diego and UC Santa Barbara entered the semiconductor field, focusing on compound semiconductors such as gallium arsenide. The program at UC San Diego moved on to embrace communication circuits. Since the 1990s, UC Irvine, UC Santa Cruz, and UC Riverside joined the research and teaching efforts in the fields of solid-state science and engineering.
- UC also contributes executive leadership to California's electronics manufacturing industry. Nearly 37% of all R&D intensive electronics manufacturing firms in the state have at least one top executive with a UC background. The UC system also contributes significantly to the industry via the involvement of UC graduates and faculty members on corporate boards and scientific advisory boards.

As a result of the increasingly symbiotic relationship between the region's research universities and the private sector over the last three decades, there has been a surge in patents granted to Silicon Valley firms. In 1980, Valley companies were granted only 250 patents. By the early 1990s, this figure increased to nearly 2,000. A decade later, it tripled to more than 6,200.[49]

Over the years, the number of patents granted to Silicon Valley companies or individuals on a per-capita basis has continued to rise. In 1994, the number of patents per 100,000 persons in the Valley stood at 114. By 2004, this number had risen to 377.[50] In that year, the Valley generated nearly 46% of all the patents granted to individuals and businesses in California, and it consistently occupies a dominant position nationally. (Figure 11)

Figure 11

Silicon Valley Patents
U.S. and California, 1995-2005

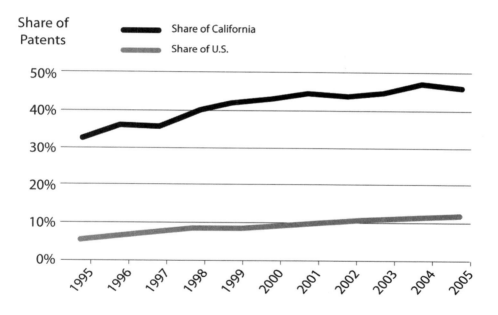

Source: U.S. Patent and Trademark Office

Of the top ten patent-generating U.S. cities in 2006, Silicon Valley was home to six; San Jose topped the list with 2,325 registered patents.[51] (Figure 12)

Increasingly, the Valley is leveraging its position on a global scale to tap the creativity in other regions and countries around the world. Co-patenting by U.S. companies with companies outside the U.S. has increased significantly in the last decade, particularly with India and China. (Figure 13)

Figure 12

Top U.S. Cities for Patents
Number of Registered Patents

1	**San Jose, California**	**2325**
2	Austin, Texas	1431
3	San Diego, California	1138
4	**Sunnyvale, California**	**1081**
5	Boise, Idaho	1072
6	**Palo Alto, California**	**922**
7	**Fremont, California**	**815**
8	Houston, Texas	800
9	**Cupertino, California**	**733**
10	**Mountain View, California**	**716**

Note: Silicon Valley cities are in bold type.

Source: U.S. Patent and Trademark Office, 2006

Figure 13

Share of Silicon Valley Patents Listing Non-U.S. Co-Inventors
1993-2006

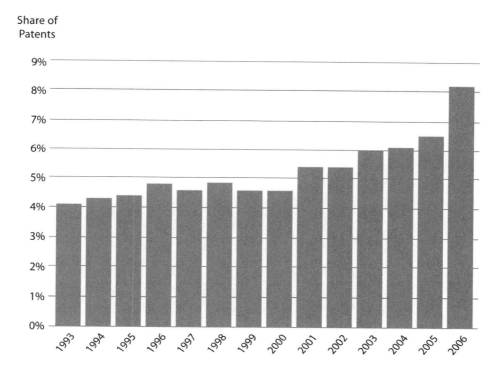

Source: U.S. Patent and Trademark Office, 2006

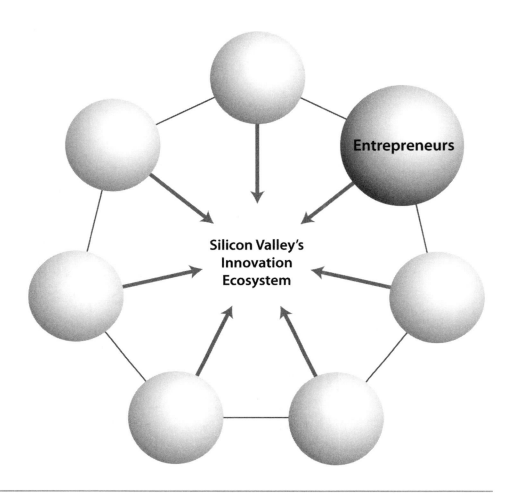

Key Element 2: Entrepreneurs

Entrepreneurs turn ideas into products. Their willingness and ability to accept the risks associated with taking an innovative product to market is influenced by the region's culture. Any region striving to develop a sustainable innovation economy needs to foster a culture of entrepreneurialism in its business community and nurture next-generation entrepreneurs in its educational institutions.

An entrepreneur is a visionary—someone with an innovative idea, a deep belief in the product around which that idea is formed, and the willingness and determination to bring that product to market.

Quite often entrepreneurs take significant personal risks and assume tremendous financial burdens. While entrepreneurs often turn to angel investors (many of whom are successful entrepreneurs themselves) or venture capital firms for money, entrepreneurs inherently have their own "skin in the game" financial burdens as they strive to bring concepts to life as technologies or products.

Profile of the High-Tech Entrepreneur

Silicon Valley has a highly entrepreneurial and risk-taking culture with a strong work ethic where bold moves, radical thinking and even the occasional honest failure are marks of distinction. It is no wonder, then, that Silicon Valley has an abundance of entrepreneurs.[52]

When we think of the stereotypical Silicon Valley entrepreneur, we most likely envision a person who is highly individualistic, blessed with brilliant ideas, endowed with talent, and who takes decisive actions that result in success. Some might say the image of the successful Valley entrepreneur is that of a daring superstar technology cowboy.

Reality, however, paints a somewhat different picture. The Entrepreneurship Center at the University of Ohio offers this profile of a typical high-tech entrepreneur:[53]

- Revolutionary: wants to change the world with great ideas and new technologies
- Wants to be a global player from the start
- Expert at developing an extensive local network

It is striking how many of the truly revolutionary companies are started, at least in part, by people who haven't done it before. Google (Brin and Page), Yahoo! (Yang and Filo), Facebook (Zuckerberg), Apple (Jobs and Wozniak), etc. When you see one of those really revolutionary companies and there's some young kid with the idea, of course, they often are linked up with one or more seasoned, experienced people—Google (Schmidt, Doerr, Moritz), Yahoo! (Moritz, Koogle), Facebook (Thiel, Breyer), Apple (Markkula). So even there you see a kind of a serial entrepreneur (or VC or executive) effect which is another form of what you're talking about.

Marc Andreasen

Source: "Serial Entrepreneurs and Today's Silicon Valley" October 29, 2007, http://blog.pmarca.com

- ✦ Highly competitive and a risk taker
- ✦ Believes in speed—time to market is critical
- ✦ Wants great financial reward but also seeks respect, influence and fame
- ✦ Ultimately wants to control the market and not just the company
- ✦ Team player in the organization and the network
- ✦ Loves to celebrate and recognize achievements and success of others
- ✦ May become an angel investor to repeat success and share expertise and ideas

Contrary to the notion that the typical Silicon Valley entrepreneur is a workaholic loner, the successful Silicon Valley entrepreneur is someone who is connected.[54] Building a new business in the fast-moving technology world is rarely a Lone Ranger task. While successful Silicon Valley entrepreneurs are often engaged in long hours of hard work, their most valuable work involves building an extensive network of managers, colleagues, investors, suppliers, and potential customers with whom they actively share ideas, business strategies, knowledge and experience, comments and critiques.

Inspiring and Supporting Entrepreneurs

Life as an entrepreneur can be quite lonely and intimidating. So it's no wonder that social and professional networks are especially important for entrepreneurs in Silicon Valley. They link inventors, engineers, entrepreneurs, investors and business professionals together. Two organizations have played important roles in the Valley's success and survival over the last fifteen years. In both good times and bad, they have inspired, trained, motivated and connected entreprenuers throughout the region.

SVASE was founded in 1995 as the Silicon Valley Association of Software Engineers, with the goal of bringing software programmers and specialists together in much the same way that computer user groups were formed in the early days of the personal computer. The well-attended meetings featured people (peers), pizza (the official food of programmers of the day) and presentations (from key people in the industry). In the wake of the Dot-Com Bust, SVASE reinvented itself as the Silicon Valley Association of Start-up Entrepreneurs, with an emphasis on helping start-up entrepreneurs from all technology sectors develop truly viable business models and find the funding to start their businesses.

SVASE now offers a wide range of educational workshops, training programs and events, including StartUp U, a "university" for new entrepreneurs; CXO Leadership Forum lunch sessions; and Learn from a Legend dinners, designed to provide entrepreneurs with feedback about their innovations and business models. For entrepreneurs looking for financing, SVASE offers numerous opportunities to make practice presentations to panels of experienced investors. In 2008, SVASE members received over $78 million in venture capital.

The second organization that fosters success in the greater Silicon Valley entrepreneurial community is the Berkeley Entrepreneurs Forum, hosted by The Lester Center for Entrepreneurship and Innovation of the Haas School of Business at UC Berkeley. For over fifteen years, the Forum has held lively monthly gatherings.

Cost-Cutting Entrepreneurs, Bootstrapping and VC Funding

The new economics of Web 2.0 (as the increasingly highly interactive use of the Web has been called) has... lowered the cost of business failures. Five years ago the cost of determining if a business was viable was $10-15 million. Today an entrepreneurial team can spend $2-3 million to find out if the business is a "go." Therefore, they can shrug off a loss and get on with the next venture. As a result, we see more experimentation and a greater number of start-ups. Bootstrapping is also increasing. That approach takes advantage of the Web's transition into a cost-effective business platform in the process of building a company through revenue and profit growth, rather than simply through outside investments. Cost-conscious entrepreneurs more often do not need VC funding as much as they did in the past.

Scott Duke Harris

Source: Mercury News, August 10, 2007

Sessions begin with an open reception for entrepreneurs and guests. Attendees wear color-coded name tags: red for entrepreneurs with red-hot technology, green for investors with green-back dollars ready to invest. It looks like speed dating, as people quickly meet, make connections, exchange cards, and move on. The formal portion of Forum events includes two-minute presentations by entrepreneurs and topical presentations by guest speakers and panels. Much of the credit for the energy and excitement of Forum sessions is owed to Jerry Engel, the Center's Executive Director, whose enthusiasm is quite contagious.

The Serial Entrepreneur

A serial entrepreneur is someone who enjoys the process of starting, growing and selling a business (hopefully at a profit), then starting and growing a new business —sometimes in a totally unrelated field. Not all entrepreneurs are successful, nor is every successful entrepreneur successful the second, third or fourth time around. In many ways, the resilience of Silicon Valley's economy is rooted in the indomitable spirit and persistence of entrepreneurs who want to change the world with their ideas despite the risks and hard work involved. Dr. Bill Musgrave, President and Chief Executive Officer of the Enterprise Network of Silicon Valley, says "no one cares about who you are—your title, how much money you have, or the color of your skin. What counts in the Valley is what you can achieve. It is the best place, bar none, for entrepreneurs to launch their dreams."

Here are a couple of examples of the Valley's serial entrepreneurs:

- From high-tech publishing to high-taste pleasure: Louis Rossetto, founder and publisher of the iconic tech-popular *Wired* magazine and wired.com, now bets he can build a successful "high-tech" chocolate business. His new venture is called TCHO, a gourmet chocolate manufacturing company with headquarters and public-friendly factory at Pier 17 in San Francisco.

Rossetto describes TCHO as "where technology meets chocolate; where Silicon Valley start-up meets San Francisco food culture."[55]

• From world-class to world-changing: Shai Agassi (no kin to tennis star Andre Agassi) is the former CEO of one of the leading software companies in the world: SAP. Born in Israel, 40-year old Agassi has now launched an effort to jump-start the electric car industry via his venture capital-backed company, Better Place. Founded in 2007 and head-quartered in Palo Alto, the company's goals are impressive: building a sustainable transportation system, and freeing the world from its dependence on oil-fueled transportation.[56] Partnering with Better Place and Renault-Nissan, Israel has committed to the widespread deployment of an electric recharge grid to power electric vehicles by 2011.[57] In addition to Israel, several other countries, including Denmark, Canada and Australia, have committed to the deployment of an electric car network. The State of California has also made a similar commitment.

Diversity: A Foundation for Global Innovation

While many of the Valley's entrepreneurs are homegrown, a 2007 survey by Duke University reported that 52.4% of Silicon Valley's start-ups had at least one foreign-born key founder.[58] This number has increased by almost 25% compared with results from a similar survey in 1999 and is substantially higher than the California average of 38.8%.

Well-known examples include Russian-born Sergey Brin, co-founder of Google; Indian-born Vinod Khosla, a co-founder of Sun Microsystems; and Jerry Yang, who co-founded Yahoo!. The study also found that "from 1995 to 2005, Indians were key founders of 15.5% of all Silicon Valley start-ups, and immigrants from China and Taiwan were key founders in 12.8%." The report's authors confirm our observations and support our sentiments, commenting, "What is clear is that immigrants have become a significant driving force in the creation of new businesses and intellectual property in the U.S.," and, as we have seen, particularly in Silicon Valley.[59]

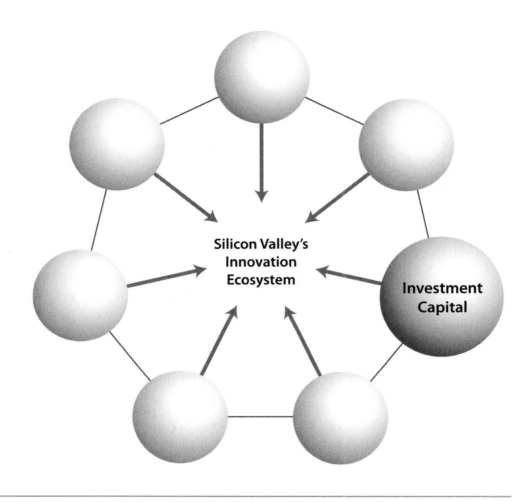

Silicon Valley's Innovation Ecosystem

Investment Capital

Key Element 3: Investment Capital

Ideas are the soul of innovation. Money is the life blood. Access to seed and early-stage money from angel investors during the start-up stages and money for expansion from venture capital firms is critical for high-tech companies. Investors support innovation by providing entrepreneurs with more than money—they provide valuable business expertise, links to personal contacts and access to social and professional networks.

Investment Capital

Typical high-tech entrepreneurs in Silicon Valley are not looking to set up lifestyle businesses—firms that support a certain lifestyle for them. Rather, they have an ambitious vision that usually involves bringing a revolutionary new product to the market and a dream of making lots of money by doing so.

Innovation Needs Cash

It's a rare high-tech company that can scale big enough and fast enough without an infusion of cash. Cash is the lifeblood of high-tech companies, particularly when a company is likely to be engaged in months or years of product development, during which phase there will be little or no income generated by sales. Like the vision itself, the entrepreneur's need for cash is significant, usually millions of dollars. The ability of an entrepreneur to raise sufficient cash to pursue his or her vision is often equally as important as the ability to envision and invent a new product or service.

When we think of cash needs on this scale, we often think of venture capital (VC), but in reality, it is the rare venture that actually gets funded by a VC firm—at least initially. More often than not, entrepreneurs invest considerable personal resources (both cash and sweat equity) before they seek additional funds. It's not uncommon for Valley entrepreneurs to take out a second mortgage on their homes and max out their credit cards in the early stages of launching a company.

At the point that personal resources are no longer sufficient to support growth, an entrepreneur must seek outside funding, typically from family and friends. This can be a mixed blessing. Certainly friends and family may be able to provide funding, but taking their money can jeopardize personal relationships and complicate future funding—particularly if promises are made to seat these

> There is a strong link between venture capital and innovation. Economists Samuel Kortum and Josh Lerner conclude from their research that a dollar's worth of venture capital can be nearly ten times more effective in the creation of patents than a dollar's worth of traditional corporate R&D.
>
> *Source: "A Special Report on Innovation," The Economist, October 13, 2007*

early investors on the company's board or non-dilution stock commitments are made. More sophisticated entrepreneurs will avoid this source of funding in favor of seeking seed and start-up money from high net-worth individuals who can not only provide cash, but also business expertise and personal connections to potential sources of the next round of funding. These high net-worth individuals, or accredited investors, as they are called by the Securities and Exchange Commission (SEC), are also often called angel investors (or business angels, in Europe).

The Key Role of Angel Investors

The presence of large and media-visible venture capital firms has often overshadowed the role of angel investors in the development of Silicon Valley's high-tech economy. VC investments are usually made in large amounts and funding events frequently attract press coverage, while angel investments are typically smaller and often occur under the media's radar. It is important to note that angel investors are the largest source of seed and start-up capital, providing early-stage funding when VCs are often not willing to take the risk. The University of New Hampshire's Center for Venture Research reports that "angels continue to be the largest source of seed and start-up capital, with 46% of 2006 angel investments in the seed and start-up stage. This preference for seed and start-up investing is followed closely by post-seed/start-up investments of 40%."[60] In a 2000 study, MIT's Entrepreneurship Center categorized angel investors in four primary groups:[61]

+ *Guardian Angels*, who bring both entrepreneurial and industry expertise. Many have been successful entrepreneurs in the same sector as the new companies they back.

- *Entrepreneur Angels*, who have experience starting companies but come from different industry sectors.
- *Operational Angels*, who bring industry experience and expertise, but generally from large, established companies, and may lack first-hand experience with the travails of a start-up.
- *Financial Angels*, who typically invest purely for the financial return.

Hans Severiens, founder of the legendary Silicon Valley Band of Angels investor group, attributes the emergence of tech-related angel investing in the Valley to the increasing use of stock options as a form of deferred compensation.[62] He notes that back in the 1970s, only the top ten or fifteen senior executives would be given stock options. But in the 1980s and particularly in the 1990s, stock options were used as a primary tool to assure that senior managers and key engineers would remain with a company long enough to achieve the firm's business goal—which often was an initial public offering (IPO) of company stock. With so many firms enjoying successes in the 1990s, many new high net-worth individuals were created. Quite a few of these newly minted qualified investors were not ready for traditional retirement, so they took their money and their expertise and went looking for start-ups to play with (and to invest in). "As the level of wealth in the Valley has increased, people that have been executives and managers at tech companies are personally investing in a lot of start-ups," says Ed Colligan, chief executive of mobile-device maker Palm, Inc. in Sunnyvale, California. "There's almost a new middle class of investors."[63]

A 2004 survey by the University of New Hampshire's Center for Venture Research found that the typical angel investor has the following characteristics:[64]

- Middle-aged and male: over 90% are male and between 40 and 60 years old
- Holds a Masters or other advanced degree and has prior start-up experience
- Has an annual income in the $100,000 to $250,000 range
- Invests 2.5 times a year at roughly $25,000 to $50,000 per deal ($130,000 total)
- Seldom invests in more than 10% of a deal

- Seeks 20% compounded per annum returns
- Expects to hold an investment for five to seven years
- Likes to invest in the technology he knows and prefers manufacturing and product companies
- Prefers to invest in fast-growing start-ups, and often prefers to take an active consulting or advisor position with a company
- Likes to invest with others and prefers to invest close by (typically within 50 to 300 miles from home)
- Invests to receive a high rate of return (e.g., 20% to 30% internal rate of return)
- Learns about deals from friends, 30% from accountants or attorneys
- Would like to see more deals, and refers deals to other private investors

Supporting Severiens' perspective on the impact of stock option wealth on the investment environment in Silicon Valley, a 2006 survey by the Federal Reserve Bank of Cleveland found that the presence of certain types of capital made it easier for a region to have successful angel investing. The survey notes that, according to participating angel investors, successful angel investing requires first-generation wealth. The reason: "Old money does not engage in angel investing because the holders of that capital are unwilling to take risks. They want to preserve the capital because, unlike first generation wealth, the people with the capital do not know how to make money. Therefore, if they lose money investing in other people's companies, they cannot make more. Moreover, first generation wealth trusts its own judgment about investments in start-up companies because it made money on the basis of that judgment."[65]

Apart from making big money, there are several non-financial reasons why Silicon Valley angels seek out and support high tech start-ups. First, being an active (rather than passive) investor allows an angel to participate in the activity, excitement and volatility that surrounds a start-up without getting overly immersed in day-to-day commitments and long work hours. Second is an often-expressed desire to "give back"—to contribute time, expertise and connections to helping start-ups in empathy for the effort involved in manifesting a vision. Also, being an active angel investor keeps a person linked into the Valley's social network—maintaining connections can have many long-term personal as well as financial benefits. And, many younger cashed-out entrepreneurs and executives see their investments in socially beneficial technologies (e.g., medical devices and biotech products) as a way for them to have a positive impact on other people's lives.[66]

The emergence of this new class of high-tech savvy investors came just at the right time to fill the increasing need of high-tech start-ups for seed and early-stage funding. In 1990, the first $100 million VC fund was established. It was followed soon thereafter by the first $500 million fund, and in the late 1990s, by the first $1 billion fund.[67] Parallel to this trend toward larger VC funds was the trend toward larger initial investments. Where start-up rounds were once $1 million, they grew to $5 million, then $10 million. As Severiens points out, a few hot start-ups were able to absorb this kind of seed investment, but most were not.[68] This led to an increasingly difficult investment environment for many early-stage companies. As a result, angel investors, often investing together as a group, stepped in to fill this growing early-stage funding gap.

Banding Together: Angel Networks

Founded in 1994, the Band of Angels is Silicon Valley's oldest angel investor organization dedicated exclusively to the funding and advising of seed-stage start-ups.[69] Members of the Band of Angels include founders of companies such as Symantec, Logitech, and National Semiconductor, as well as senior executives from Sun Microsystems, Hewlett Packard, and Intuit. The group made fourteen new deals between mid-2006 to mid-2007 plus another thirteen follow-on investments. Since its founding, the Band has completed 201 deals. Nine of the Band's portfolio companies have gone IPO while 43 have been sold at a profit. What this says is that even with such a savvy and experienced group of investors selecting the deals, only 25% of the Band's investments have been profitable.

Thus, it is not surprising that each deal is expected to make a 10-times return on investment, with the assumption that most deals will ultimately fail to be profitable.[70]

The Angel Capital Association reports that angel investors often do ten times as many deals as venture capital firms, but the average success rate is roughly the same.[71] This statistic reflects the risks associated with early-stage investment. Still, the attraction of angel investing remains strong despite the high level of risk, as the potential payoff can make it worthwhile. Don Dodge, Director of Business Development for Microsoft's Emerging Business Team and commentator on angel investing, notes, "One big winner can cover lots of losers."[72]

Investments by the Band of Angels range between $300,000 and $750,000, but the group often leads a syndication of $2 million to $3 million. The largest investment ever made by the Band was $3.3 million. Looking at a profile of the Band's investing, we see that the group's portfolio companies are distributed in a wide range of high-tech fields. (Figure 14) Mirroring a profile of venture capital investing by sector, the Band's top investment shares go to software, life science/biotech, and semiconductors. The software and biotech sectors have consistently received the largest shares of venture capital investments nationwide.[73]

Given that angel investors are private individuals, their activities are often difficult to measure and track. The University of New Hampshire's Center for Venture Research is a primary source of data on angel investing in the United States. The Center reports that investing by angel investors totaled $11.9 billion (24,000 deals) in the first half of 2007, down 6% in the first half of 2007 over the same period a year ago. The number of active angel investors increased 8% to 140,000 people this year. The health care and medical devices sector received the largest share (22%) of angel funding in the first half of 2007. The software sector was second with a 14% share, followed by biotech at 10% in third place.[74]

The Dot-Com Bust drove many angel investors out of the market. Those who continued to at least consider early-stage investing after the 2000-2001 Dot-Com Bust have been understandably much more cautious and have found

Figure 14

Investment Profile of Silicon Valley's Band of Angels

Investments by Industry Since 1995

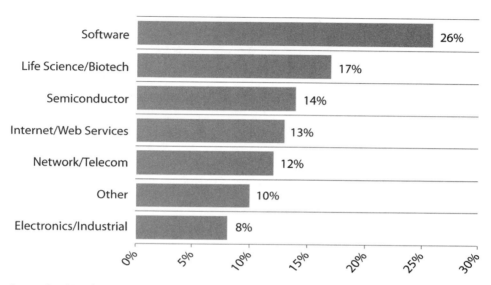

Source: BandAngels.com

strength in numbers. Since the formation of the Band of Angels, numerous other angel investor groups have been established. At a practical level, the process of finding and vetting deals is time-consuming for an individual acting alone. Working together, an angel network leverages the team's collective experience, increases deal flow and distributes the burden of due diligence. By sharing responsibilities and due diligence tasks among members, angel groups are more likely to make better investments.

In addition, angel groups typically are better at positioning their portfolio companies for next-stage financing from venture capital firms. Historically there has been some competitiveness between angels and VCs, particularly during the Dot-Com Boom, when investors of all kinds were trying to be first in on hot deals.[75] These days, there is more respect for angel investors within the VC community, especially given that VCs are much more involved these days in much larger deals. They increasingly appreciate the important role that angel investors play in supporting innovative entrepreneurs at the very early stages, particularly as mentors.

A classic example of the mentor relationship is exemplified by Robert Noyce, one of the Traitorous Eight who founded Fairchild Semiconductor and subsequently served as its general manager, and Steve Jobs, co-founder of Apple Computer and of Pixar, Inc. Noyce was a valued resource whom Jobs deliberately sought out and befriended.[76] In fact, Noyce and wife Ann Bowers would let Jobs occasionally sleep on their couch.[77] Noyce became one of the first investors in Apple. Later, in the Valley's spirit of entrepreneurial mentoring, two young Stanford grad students, Larry Page and Sergey Brin, specifically sought out Jobs as a mentor in the early days of their start-up: Google.

Venture Capital: Overview

As we've seen, both angel and venture capital investments have fueled Silicon Valley's innovation economy. While seed- and early-stage investments by angels are often critical for launch, angels are rarely able to provide funding sufficient to take companies to the next level and beyond. This is the role that venture capital firms play. Where the typical angel investment today is under $250,000 and, even when syndicated, investments rarely reach beyond a couple of million dollars, the average Silicon Valley VC deal in the second quarter of 2007 was $8.7 million. As one venture capitalist put it, it takes almost as much time and effort to find, screen and close a smaller angel deal as it does a large VC deal. With the steadily increasing size of VC funds, fund managers have little time for small deals.[78]

As defined by the National Venture Capital Association, venture capital firms are "pools of capital, typically organized as a limited partnership, that invest in companies that represent the opportunity for a high rate of return within five to seven years. Far from being simply passive financiers, venture capitalists foster growth in companies through their involvement in the management, strategic marketing and planning of their investee companies. They are entrepreneurs first and financiers second."[79] This is particularly true in Silicon Valley. Until the mid-1970s, individual investors were the "archetypal venture investor."[80] Once federal regulations were changed in order to allow pension funds to make investments in alternative asset classes such as venture capital funds, venture capital firms began to play an increasingly important role in financing larger and later-stage deals.

In 1978, the federal government removed various restrictions on institutional investing and slashed the capital gains tax rate from 49.5% to 28%.[81] Still, after a seven-year stock market slump in the early 1970s, venture capital fund-raising was tough. In 1978, the industry raised approximately $750,000 nationwide.[82] In 1980, this figure had grown to $600 million. By 1987, VC fund-raising soared to almost $4 billion. In 2006, U.S. venture capital funds raised over $24.7 billion.

Venture Capital: Silicon Valley

The first Bay Area venture-backed start-up is generally considered to be Fairchild Semiconductor, funded in 1959 by Venrock Associates, a pioneering VC firm formed in 1969 by Laurence Rockefeller, son of John D. Rockefeller. Since then, Venrock has invested more than $1.8 billion in more than 400 companies, including Apple Computer, Intel, and 3Com.[83] Silicon Valley now has one of the largest concentrations of venture capitalists in the world and receives the largest share of venture capital investment of any region in the U.S. and, in fact, of any region in the world. (Figures 15, 16, and 17) Legendary Sand Hill Road in Menlo Park is home to some of the best known high-tech VC firms, including Kleiner, Perkins, Caufield & Byers. Unlike these other financial centers, Sand Hill Road is within an easy drive of many high-tech headquarters, start-up companies and the Stanford University campus. The location provides them with convenient access to deal opportunities—after all, VCs like to keep a close eye on their investments.

In a 2000 study of the Valley's social networks, Emilio Castilla, et al, noted that "in Silicon Valley, venture capitalists play more than their conventional roles; they influence the structure and future development of their client companies."[84]

Figure 15

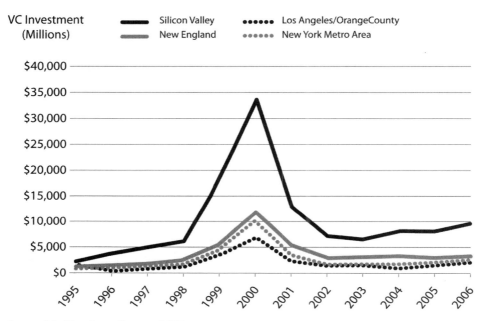

Venture Capital Investments by Region
1995-2006

Source: PriceWaterhouseCoopers, 2007

VCs protect their investments by taking seats on a portfolio company's board of directors. If need be, they will sometimes reorganize the board or remove the original founders from the company, as happened with Cisco Systems and Silicon Graphics.[85]

As one would expect, an active VC will be involved with more than one portfolio company at a time. A 2006 survey by the National Venture Capital Association and Dow Jones VentureOne reported that VCs believe the ideal average number of boards an active VC should be engaged with is 4.6 for early-stage companies, 5.5 for later-stage companies. CEO's prefer that their venture capitalists limit their early-stage board seats to an average of 4.0 and their later-stage board seats to 4.6.

Figure 16

Venture Capital Investment by Regional Share and Deals
Q1-2007

Region	Investment ($ Million)	Percentage of Total	Deals
Silicon Valley	$2,165	30.89%	244
New England	$976	13.83%	95
San Diego	$686	9.73%	44
Southeast	$579	8.21%	54
Los Angeles/Orange County	$526	7.45%	40
Northwest	$457	6.47%	42
New York Metro	$360	5.11%	48
Midwest	$280	3.98%	43
Texas	$278	3.93%	34
Philadelphia	$205	2.90%	34

Source: PricewaterhouseCoopers, 2007

Geographically, Bay Area VCs averaged the highest number of board seats at five per VC.[86] The survey notes that there are many Silicon Valley VCs who sit on more than ten boards.

In 2006, Silicon Valley-based companies received nearly one-third of all domestic VC investments.[87] In Q1-2007, total VC investments in Silicon Valley were more than 2.2 times the level of the next "most popular" region, New England—over 30% of the total investments and deals nationwide. In Q2-2007, VCs invested $2.52 billion (290 deals) in Silicon Valley companies, up from $2.37 billion (269 deals) in Q1. Sector investments included $283 million in 29 semiconductor deals; $248 million in 17 networking and equipment deals; $167 million in 15 industrial or energy deals, $137 in telecom deals, and $79 million in six computer deals.

Figure 17

VC Investments in Silicon Valley

Quarterly Data, Q1-1998 to Q1-2007

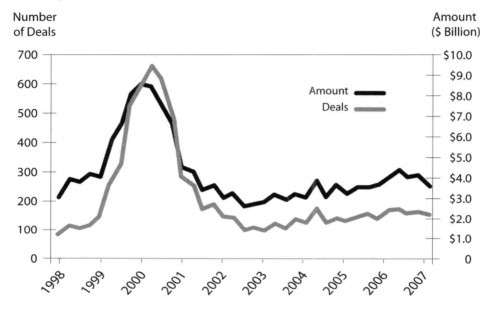

Source: PriceWaterhouseCoopers, 2007

The Globalization of Venture Capital

Silicon Valley's venture capital profile is changing as the region's investors become much more globally interconnected. Today Valley VCs are holding more globalized portfolios in order to enhance opportunities for international collaboration for jointly developing intellectual property (co-patenting) as well as to expand their global reach in terms of innovation.

Over the last decade, there has been a shift in the flow of VC investments in and out of Silicon Valley. In 2007, the Joint Venture: Silicon Valley Network (JVSVN) reported that between 2001 and 2006 Silicon Valley VCs significantly increased their investments in Asia, the U.K. and Israel, while decreasing investments in Germany, Taiwan and Singapore. Shanghai, Beijing and Seoul showed the largest gains by city or region. At the same time, Chinese investments into Silicon Valley remained less than from any of the other primary investing country. Investors from the U.K., Israel and Taiwan meanwhile increased their investments in Silicon Valley companies. The Joint Venture report concludes that "patterns of capital flows suggest interconnectivity between regions not only in terms of investment but likely also in terms of talent and technology."[88] The Valley clearly benefits greatly from these international connections. (Figure 18)

Figure 18

Investments Into and Out of Silicon Valley
2001–2006

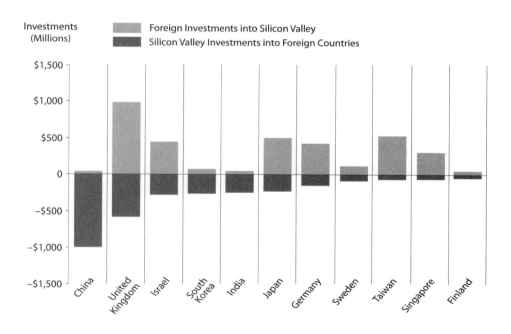

Source: Joint Venture: Silicon Valley Network, 2007

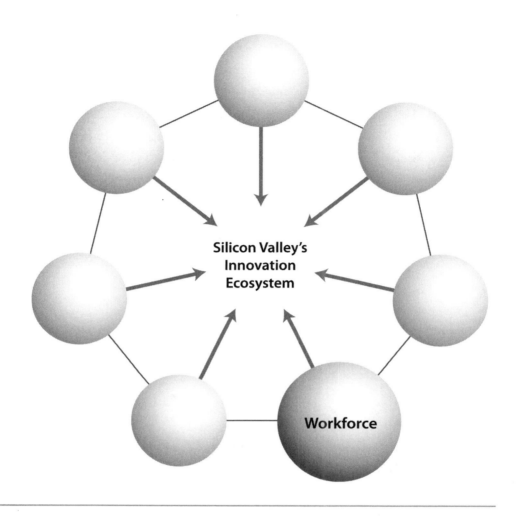

Silicon Valley's Innovation Ecosystem

Workforce

Key Element 4: Workforce

An innovation ecosystem needs a large pool of talented workers to fill the myriad of positions created by new start-ups, expanding companies and well-established firms. Workforce diversity brings new ideas, new connections and access to global resources. Smart companies understand that competition for talented workers is stiff and that perks are often an important factor in keeping employees happy and loyal.

There is no question that Silicon Valley has a world-class workforce. Some of the best and brightest scientists, engineers and technical experts from around the world work for Valley firms. The Valley's workforce is the foundation of its high level of productivity.

The Valley's level of value added per employee tops the nation and has been steadily increasing.[89] In 2006, the U.S. average value added per employee finally reached the level that Silicon Valley employees attained almost a decade earlier. (Figure 19) Output from the Valley's workforce now stands 20% higher than the national average. Between 2005 and 2006, the annual value added per employee in the Valley increased at a rate of 4.1%, more than double the U.S. rate of 1.9% for the same period.[90]

Workforce Diversity

Looking at Santa Clara County's migration patterns over the last decade, we can see the changing face of Silicon Valley. International in-migration (people arriving from outside the U.S.) over the last decade has been responsible for adding approximately 20,000 to 22,000 new immigrants per year to the Valley. (Figure 20) Almost two-thirds of the immigrant workers were of Asian background, mostly Chinese and Indians. The Asian Indian community grew by 43% (8,000 people) between 2000 and 2005—the largest percentage gain for an ethnic group. The Chinese community grew by 24% (11,000).

Skilled immigrants are a very important part of the Valley's workforce. Silicon Valley's talented engineers, programmers, designers and executives tend to be of very diverse backgrounds and the share of talent coming from other countries has also been increasing. Immigrants accounted for more than 30% of the engineering workforce in the 1980s and 1990s. In 2007, the Joint Venture: Silicon

Figure 19

Silicon Valley's Productive Workforce
U.S. and Silicon Valley, 1996-2006

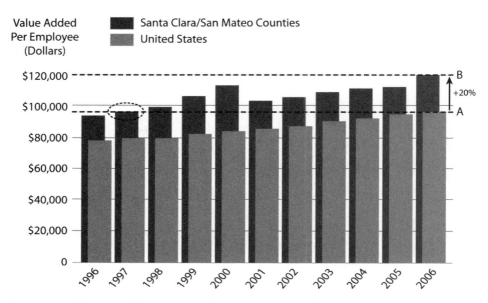

Note: Average U.S. productivity has consistently lagged a decade behind Silicon Valley.

Source: Joint Venture: Silicon Valley Network, 2007

Valley Network reported, "Over half of the region's science and engineering (S&E) talent was born abroad. In 2000, this group constituted 49%, and by 2005, it expanded to 55% of the region's science and engineering occupations. Foreign-born talent in Silicon Valley represents roughly three times the national shares in S&E and in all occupations."[91]

The inflow of talent from India and China has been particularly remarkable. The 2000 U.S. census reveals that 13% of the scientists and engineers in the Valley were born in India, 9% were born in China—a total of 22%. This proportion of Asian scientists and engineers working in the Valley was twice that of Boston, Seattle, or Austin. Today, when Valley technologists say that the Valley is "built on ICs," they are no longer talking about "integrated circuits," they are

Figure 20

Net Domestic and International Migration
Santa Clara County, 1997-2007

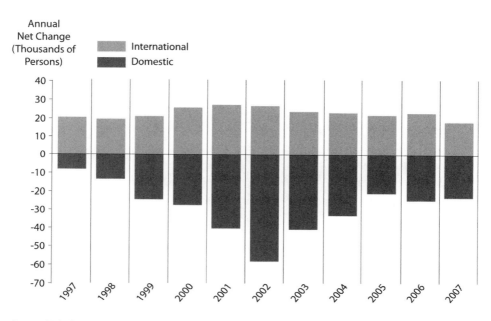

Source: U.S. Census Bureau

referring to Indian and Chinese engineers.[92] One reason for this extraordinary inflow of talent is that unlike many other developing countries, India and China tailor the education of their students toward admission to U.S. universities. Other reasons include Silicon Valley's culture of meritocracy, high acceptance of diversity, and the well-developed social and professional networks of Chinese and Indian residents in Silicon Valley.

Retaining Talent: The Perk Wars

In such a hypercompetitive environment, Valley businesses are challenged to retain their talented employees. With the arrival of the Dot-Com Boom of the late 1990s, Valley start-ups went all out to lure top technical talent with perks such as flexible workdays, world-class cuisine in company cafeterias, and the freedom to have unconventional office decors such as lava lamps and empty beer can sculptures—all this plus stock options that could turn a young engineer or programmer into a multi-millionaire overnight. In the heyday of the Dot-Com Boom, well-established Silicon Valley companies were forced into a perks war—offering more and more perquisites and workplace benefits—in order to stay competitive with start-ups. So they began offering all of the same perks (except for stock options).[93] The perks war ended with the Dot-Com Bust of 2001, when thousands of Valley employees were laid off as hundreds of start-ups went broke.[94]

With the Valley's economy fully recovered from the bust and facing fierce worldwide competition for top talent, the perks wars are back with a vengeance. Today, the range of the perks offered is more impressive than ever: on-site gyms, car mechanics, cafeterias, childcare, snack and coffee bars, laundry services, dentists, massages, vacation plans, free sessions with psychological counselors, yoga lessons, and weight loss programs. (Figure 21) While some of these perks are being trimmed back due to the crisis of 2008-09, they still remain a vibrant aspect of Silicon Valley's workplace.

These on-site perks are intended to make workers feel valued and recognized, and to help them take care of their health and fitness needs. Most importantly for the firm, these perks are designed to encourage employees to work more productively, sometimes for long hours. Ten- to fifteen-hour workdays are not unusual in the Valley.

Looking over a 2007 study that surveyed 50 best places to work in the Valley,[95] including iconic names such as Google, Yahoo!, Applied Materials, Cisco Systems, eBay, Hewlett Packard and Intel, vacation days were also part of the perks package. Out of the 50 firms surveyed, 44 clearly stated the number of annual vacation days they offer to employees. The vacation days range from 10 to 30 days a year (the latter offered by Sun Microsystems). Twenty-three of the firms provide 15 days of vacation a year.

Figure 21

A Sampling of Company Perks

Adobe Systems, San Jose
Free snacks and drinks, on-site fitness center, rooftop basketball and bocce ball courts, free one-year Caltrain pass, ergonomic office furniture, and free downtown San Jose parking anytime

BEA Systems, San Jose
On-site haircuts, on-site car washes, adoption assistance, infertility treatment assistance, smoking cessation program, fitness club discounts, concierge services, and discount tickets

Cadence Design Systems, San Jose
Ergonomic evaluation and office set-up; minimum 10 weeks paid maternity leave, community volunteer leave program for disaster relief, on-site fitness center, dental care, food service, and oil changes

Cisco Systems, San Jose
Free snacks and beverages; on-site health screenings, fitness centers, childcare center, banking services, dry cleaning, car-care services, haircuts, dental care, and breastfeeding program

Electronic Arts, Redwood City
Employee enrichment classes, discounts on video game systems; private pre-release movie screenings; print and media library, on-site massage, game room, dental care, and chiropractic services

LSI Logic, Milpitas
Adoption assistance, on-site ATM machines, cell phone services discounts, patent award program, and finance seminars

National Semiconductor Corp., Santa Clara
On-site credit union, fitness center, café, dry cleaning, 14-acre employee park (with lake, par course, soccer and softball fields, and picnic areas), dollar-for-dollar match for cash donations to charities and education

Network Appliances, Sunnyvale
Autism benefit, volunteer time-off program, on-site 24/7 fitness center, special carpool parking, bike lockers, farmers' markets, on-site dental care, sports leagues, on-site massages, hypnotherapy and meditation

SanDisk Corporation, Milpitas
On-site 24/7 gym with classes and personal training, café, book fairs, charitable giving program, on-site dry cleaning, haircuts, car wash, and homework assistance for school-age children

SRI International, Menlo Park
On-site massage, dry cleaning, shoe repair, photo service, dental service, auto detailing, transit assistance, and emergency backup childcare

Source: San Jose Magazine, *November 2007*

Google's extraordinary working conditions and benefits provide us with a glimpse of the extent to which top Valley companies are making sure that they are able to hire and retain the best talent. During a recent visit to the company's Mountain View campus, the following perks were standouts:[96]

+ Gourmet cafeterias for the employees
+ Kiosks that dispense free tea, coffee, latte, soft drinks and snacks in all the campus buildings
+ Free gyms complete with access to trainers and massages
+ Free laundry, dry cleaning, haircuts, a car wash, commuting buses, and so on.

Google is like a workplace paradise. According to Google CEO Eric Schmidt and Berkeley Professor Hal Varian, a consultant for Google, the rationale for generous perks is that knowledge workers are paid to be effective. Employees do not just work from 8 AM to 5 PM in the Valley. Intelligent businesses try to eliminate everything that interferes with a worker's creativity and effectiveness.[97]

Another fundamental business principle at Google is teamwork. Most projects in the company are team-led. Most offices are shared, with an abundance of small- to mid-sized conference rooms where larger teams can confer without disturbing their neighbors. This translates into fewer e-mails, faster communication, and easier working coordination. Lone Rangers apparently do not survive at Google; they are weeded out in initial job interviews.

Encouragement of creativity is also an important value of the company. Engineers can spend one day a week on projects of their choice with some oversight by management. Some of Google's best ideas are products of this creative allocation of time.

Many of the workplace practices at Google are also followed to some extent at many of the large well-established companies in Silicon Valley. Companies such as Google can always attract the best and brightest talent because they are magnets for science and technology graduates. The big question is: Can our great universities continue to produce a steady flow of highly talented scientists and engineers in the coming years?

Retraining Talent: Greening the Valley's Workforce

Just as the survival and sustainability of the Valley's economy depends on the inspirations of its inventors and dedication of its entrepreneurs, it also depends on the adaptability and skills of its workforce. The Valley's workforce is not complacent about staying on top of its game. With the fast pace of changes in technology, market opportunities and business practices, the Valley's workforce must be ready to learn new skills, change work practices, and interact in new ways.

Once again, the Valley's workforce is being challenged by change, this time by the economic crisis of 2008-2009. To survive, the Valley's businesses and industries are repositioning themselves to catch the wave of the cleantech revolution. At the high end of the job spectrum, the Valley continues to need and attract highly educated and skilled technical workers. At the middle and lower levels, though, Valley workers are scrambling to burnish their green credentials. To assist them, Valley leaders are increasing their commitment to workforce development, strategically targeting skills needed for the growing number of green jobs. The Silicon Valley Workforce Investment Board expects its training and job counseling budget to grow by more than 50% in 2009. And this is only the beginning. More resources are likely to be allocated in this area with the passage of President Obama's economic stimulus plan.

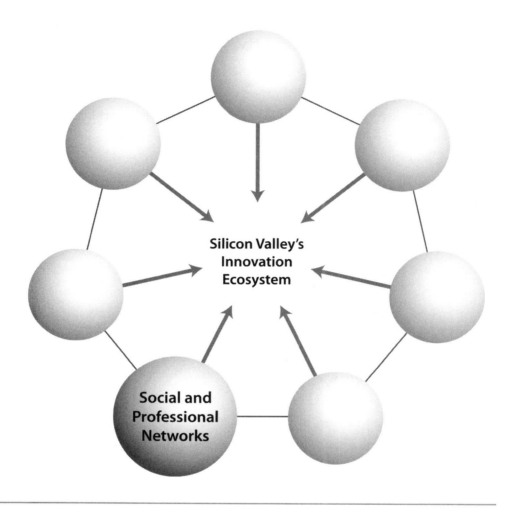

Silicon Valley's
Innovation
Ecosystem

Social and
Professional
Networks

Key Element 5: Social and Professional Networks

Inflow of top-flight talent is critical to the Valley's success. Formal and informal social and professional networks are important interconnected webs linking inventors, innovators, entrepreneurs, investors, engineers, as well as lawyers, bankers, accountants and other professionals. These networks are channels for sharing ideas and information and connecting with resources, money, talent, markets, partners, suppliers and customers worldwide. They also build community and a support system for the vast diversity of people who come from all over the world and work in the Valley.

A nother fundamental element of the Valley's success and core strength of the region is its interconnected and highly collaborative business environment, which allows Valley companies to draw on each other's strengths and facilitates the rapid exchange of information.[98] Through the Valley's formal and informal business, professional and social networks, valuable exchanges occur. Ideas, contacts and resources are shared widely among university researchers, venture capitalists, entrepreneurs, engineers and supporting professionals. The value of networks in a regional economy has been well documented. In her 1994 report on the Valley's culture,[99] AnnaLee Saxenian found that networks not only foster the free flow of ideas, they also support the formation, financing and functioning of Valley companies. The result of this symbiotic flow has been higher productivity and greater economic prosperity.[100]

The importance of Silicon Valley's social networks cannot be overstated. For a region like Silicon Valley, social networks play three very important roles:

+ Networks help shape and define the labor market, particularly where there is a highly mobile workforce.
+ Networks help define the region's power and influence structure.
+ Networks play an important role in the transformation of innovations into successful products and services.

Defining the Region's Labor Market

Valley firms benefit from the social networks of employees as they help in attracting appropriate workers to the company. The social connections of workers are seen as valuable resources for the company. In many cases, companies pay monetary bonuses to employees who provide successful referrals. Silicon Valley experience suggests that workers hired via the social network stay with a company longer and do a better job than those who are hired through other channels.

It is interesting to note that one explanation for the high degree of mobility among engineers in the Valley is that they develop personal loyalties to enhancing a particular technology rather than to a particular firm. Thus, engineers tend to move between firms via the vast network shaped predominantly by personal and technological loyalty.

Defining the Region's Power and Structure

In Silicon Valley, networks are also an important source of influence and wield the power to get things done. For example, VCs and attorneys do more than serve their conventional financial and legal roles; they shape the future direction of the companies they help. Lawyers are not only legal advisers: They are also dealmakers as they help connect start-up businesses to investors and useful sources of knowledge and information.

Investors in turn act not only as financiers. They also act as management consultants and help recruit high-level talent. They often intervene in the operations of the start-up firms in which they invest. Experienced VCs have a wealth of practical experience in running high-tech businesses and can be of great help to engineers and scientists who may be inexperienced in management and marketing.

Transforming Innovations to Products

Social networks are important laboratories where ideas are transformed into products. Silicon Valley firms are not just isolated islands floating in a sea of brutal competition. Social networks provide much-needed information and intelligence for understanding changing markets and changing technologies via informal communications among employees of different firms, membership in various professional organizations, and university affiliations. The high level of labor mobility between small firms and large firms, and between firms and industries, further enhances workers' knowledge and skills. This process helps transform ideas into innovations, and innovations into successful products and services.

It is not just that Silicon Valley has social and professional networks, but that it also has the kind of networks that create valuable interconnections among people and among companies. They foster linkages to research universities and think tanks in the region and beyond. This interconnectedness helps entrepreneurs create start-ups by providing access to critical resources. The Valley's networks connect immigrant entrepreneurs from India, China, Iran, Israel, France, and elsewhere with the regional labor market and social organizations. At the other end of the spectrum, the Valley's networks also connect VCs in India, Taiwan, Israel, etc., with start-ups in Silicon Valley. These networks and resulting interconnections are unique and are not as prevalent in other high-tech regions to the extent that we see them in the Valley. This attribute alone sets Silicon Valley profoundly apart from other high-tech regions in the United States and around the world.

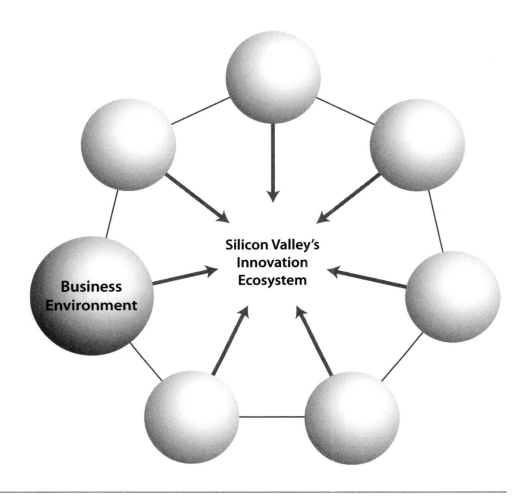

Silicon Valley's
Innovation
Ecosystem

Business
Environment

Key Element 6: Business Environment

The business environment of a place affects corporate viability and the ability of a region to compete in the global marketplace. It also affects the fundamental ecology of innovation of the region. Without favorable business conditions, companies operate at a competitive disadvantage and the relationships that exist between the various components of the ecology are out of balance.

Business Environment

I n a natural environment, just as latitude, temperature, moisture, air pressure, topography, and many other factors cause the weather of a region to be in constant flux, a myriad of interacting factors in an economic ecosystem is constantly changing. At any given time, the business climate represents a snapshot of the overall business and economic environment at that time. In addition to macro factors (the social framework and political structure, physical and economic infrastructures, population profile, etc.), a region's day-to-day business climate is influenced by other constantly changing factors. Among these are the prevailing attitudes of the local, regional and state governments and financial institutions toward business and economic activity, the current taxation regimen, costs of living, and so forth.

Assessing the Valley's Business Environment

The best tool for assessing the state of Silicon Valley's business environment is provided by the annual business climate survey conducted by the Silicon Valley Leadership Group (SVLG), a public policy advocacy group that in 2008 represented nearly 200 major Valley corporations. A comparison of survey results for 2004 and 2007 provides us with a perspective of how business climate issues shift as economic, demographic, technological, and political factors change over time.

In December 2004, SVLG distributed its first annual survey on business and living conditions in the Valley.[101] SVLG sent questionnaires to CEOs and senior corporate officers about issues of concern as well as policy recommendations for possible remedies. Over the following six weeks, SVLG received 114 replies, a respectable response rate of 59%. Here is a sampling of the results of the 2004 survey.

The top five business challenges were:
+ High housing costs for employees (68%)[102]
+ Business regulations (65%)
+ Workers' compensation costs (53%)
+ Health care costs (47%)
+ Worker recruitment and retention costs (42%)

The top policy actions suggested in the survey were as follows:
+ Help create more affordable housing (55%)
+ Reduce rates for workers' compensation costs (54%)
+ Continue the research and development tax credit (43%)
+ Invest in traffic relief and transportation improvements (43%)
+ Avoid split roll tax where business and residential property values are assessed differently (41%)
+ Limit frivolous lawsuits (40%)

In contrast, here is a summary of the 2007 SVLG survey results (115 companies responding).[103]

Housing

An unprecedented 99% of the CEOs surveyed indicated that the high cost of housing was the number-one challenge for Silicon Valley workers. In parallel, nearly 84% of the employers stated that housing cost was the top business concern in 2007. Carl Guardino, the CEO of the SVLG, stated that recruiting and retention of talented workers has been a serious challenge for businesses for many years in the Valley because of high housing costs.

Traffic Congestion

It was not a surprise to find that traffic congestion and transportation problems ranked second in the 2007 survey in a booming region that has never had adequate public transit or transit-based housing. It is well known even to casual visitors, not just daily commuters, that the arterial roads to Silicon Valley are always congested, and that they are a nightmare during rush hour. Only when the Valley's economy is down (as it was during the Dot-Com Bust of 2001) is congestion alleviated somewhat.

The good news is that in the November 2008 election, the Valley's voters approved a bond issue to finance an extension of BART (Bay Area Rapid Transit). This project will connect the Valley to the rest of the Bay Area via this efficient regional electric rail system. Connecting Silicon Valley to San Francisco and the East Bay will enhance the economy and the quality of life of the entire San Francisco region, not just the Valley. It is quite amazing that the Valley's electorate did not agree to join the BART system at its inception a few decades ago. Of course, it is now better late than never.

Health Care

Rising health care costs placed third in the survey. Health care is a serious cost of living problem for Valley workers as well as for Valley employers.

Other Challenges

Other challenges cited by the Valley's top executives were elementary through high school education (35%), high taxes (35%), energy costs (15%), childcare costs (13%), higher education (9%) and other (1%). Other challenges cited by Valley employers included business regulations (44%), immigration (37%), high taxes (34%), worker skill level and appropriate workforce (32%), energy costs (27%), and workers' compensation costs (23%).

Recommendations

CEO recommendations for the top five actions that the state of California could take in order to improve the business climate included these: help create more affordable housing (65%); improve elementary through high school education (59%); invest in traffic relief and transportation improvements (59%); continue the state's Research and Development tax credit (45%); and limit frivolous lawsuits (35%).

Several observations are in order. When we compare the top five issues in the 2004 and 2007 surveys, we find that the high cost of housing is the consistent number-one concern. The outlook may improve as we look ahead. Home prices in the Valley have fallen dramatically in recent months and are expected to continue to do so well into mid-2009. In November 2008, the typical single-family home in the Valley sold for $483,000, down a record 39.5% from November 2007.[104] The last time Santa Clara County's median price was under the half-million-dollar mark was in March 2003, when the figure hit $480,000.[105] The current price drop will undoubtedly soften concern about housing costs in the Valley.

Second, health care continues to be a nagging concern. This is not just a Silicon Valley problem, it is a state and national problem, as our health care system is inefficient as well as inadequate. On a per-capita basis, U.S. spending on health care is among the highest in the world and results are not at all commensurate with the extent of spending. Systemic reform nationally is necessary before real health-care cost reductions can take place in the Valley or anywhere else in the U.S. There is hope that the Obama Administration will be able to make improvements in our health care system, though this is a daunting challenge.

The 2008 SVLG survey, to be released in the spring of 2009, will shed new light on the Valley's changing business environment. In light of the current economic and financial crisis, it will be most surprising if the Valley's economy is not cited as the number-one problem in the results, replacing housing costs.

It is important to realize that all the different concerns expressed in these surveys have adverse impacts on the vitality of the innovation ecosystem of Silicon Valley. While business climate issues may change rank from year to year, that does not mean that the underlying problems have gone away—most continue to sap the viability of the innovation ecosystem and need attention. Solving these problems takes leadership on the part of local, state, and federal governments. They need to respond appropriately with new regulations and legislation that address problems such as housing, transportation, and education; and with public sector investments in infrastructure projects such as roads, bridges, airports, and public transportation.

Tracking and Supporting the Innovation Economy

Most regions in the U.S. have economic development agencies, organizations, and associations. These are often public-private partnerships set up to enhance and support the economic vitality of the region. Unlike chambers of commerce that exist primarily to help individual businesses grow and succeed, economic development organizations are dedicated to the vitality of the entire local or regional economy. The Joint Venture: Silicon Valley Network (JVSVN, www.JointVenture.org) is a prime example of such a public-private partnership. It does a world-class job of assuring that the Valley's business environment remains supportive of the other key elements of the region's innovation ecosystem.

Founded in 1992 by a cross-section of the stakeholders in maintaining Silicon Valley as the world's most dynamic economic ecosystem, the JVSVN is a highly respected economic development organization. Representatives from business, labor, government, the region's universities, and the non-profit sector work aggressively on a wide range of initiatives to assure that the Valley maintains its competitive edge. These initiatives include preparing a comprehensive disaster preparedness plan and encouraging the deployment of next-generation communication infrastructures. JVSVN's Wireless Silicon Valley initiative is pushing for blanketing the region with high-speed wireless Internet accessibility. Its motto is "When you no longer need a wire, doors open for Silicon Valley's wireless entrepreneurs." JVSVN is funded by a roster of sponsors that include icons of the high-tech world such as AT&T, Cisco Systems, AMD, Adobe, Google, Hewlett-Packard, and Microsoft.

The JVSVN annually publishes the *Silicon Valley Index*, one of the most useful reports designed to inform civic leaders, the business community and the general public about the current state of affairs in the Valley. This forward-looking report not only presents the standard set of indicator data on employment, housing, and population growth, but also analyzes trends in venture capital investing, clean technology and the Valley's emerging green economy. As the presenters of the *Index* noted in their release of the 2009 edition, "Regions that want to thrive first of all need a means to assess themselves." Measurement and analysis of progress is essential to the development of sound public and private policies—and JVSVN's *Index* is an excellent model.

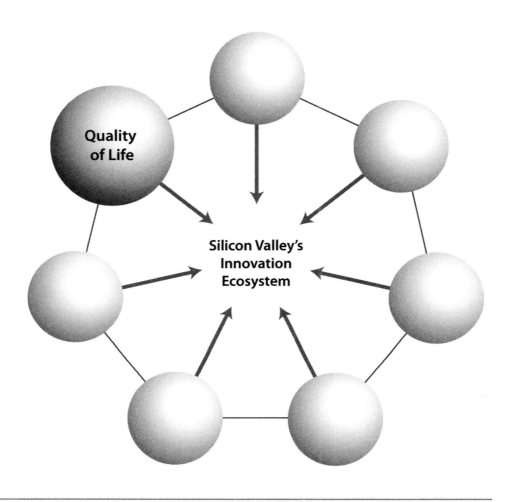

Key Element 7: Quality of Life

Innovators, entrepreneurs, investors, engineers and the many others who contribute
to the success of a company all appreciate, enjoy and often demand a high quality of
life for themselves and their families. Good schools and easy access to recreational
opportunities and cultural venues are important factors for defining the quality of life
in a region.

The importance of Silicon Valley's quality of life is not to be overlooked: It ranks as a key element of the region's innovation ecosystem. It is important to note that the quality of life enjoyed by residents of Silicon Valley extends beyond the 29 cities of the South Bay region. To a great extent, the Valley's overall quality of life is defined by the attributes of the eleven counties of the greater San Francisco Bay Area.

Mediterranean Climate

The Bay Area and Silicon Valley's climate is similar to a Mediterranean climate with mild, moderately wet winters and dry summers. It is strongly influenced by the cooling currents of the Pacific Ocean and San Francisco Bay, which tend to moderate temperature swings and produce a remarkably mild climate with little seasonal temperature variation. Due to its sharp topography and maritime influences, the region exhibits several microclimates. The high coastal hills to the west are responsible for significant variance in annual rainfall between different parts of the Valley and serve as a barrier to the coastal fog that often blankets San Francisco to the north. One can literally relax on a beautiful beach one day and ski in the Sierras the next. No snow to shovel, no humidity, no hurricanes.

Quality Living Environment

Location matters, even in the footloose digital age. It matters a great deal because knowledge workers of Silicon Valley are a fairly sophisticated bunch. They value access to universities as well as quality schools for their kids. They appreciate cultural richness and civic virtue. They expect choice in housing and transportation, diversity of employment opportunities, a good physical environment, good food, and good coffee (an important fuel for the innovation

> There are three things that have made the Bay Area a leader in technological innovation: great universities that provide strong educational resources; an availability of investment capital, and a great location that draws a diversity of talented people, including immigrants. And that's not even mentioning that we have the best climate and environment anywhere. These are the area's birthrights and we must challenge anything that threatens them.
>
> **Dr. Robert Morris**
> **BASIC Chairman Emeritus and Vice President, Assets Innovation, IBM Global Services**
>
> *Source: Bay Area Science and Innovation Consortium*

economy). In light of this, it is not surprising that even in the midst of the Dot-Com crisis era, one of the worst times for Silicon Valley and the Bay Area, *Fortune* magazine ranked San Francisco and San Jose in the top five best cities for business.[106] The criteria for selecting these cities included the business environment, the cost of doing business, the quality of the local workforce and the quality of life.

Places like San Francisco provide enormous benefits for the various communities of the Bay Area, including Silicon Valley, the rapidly growing tech-savvy East Bay, and Marin County. A culturally and aesthetically rich city like San Francisco becomes a catalyst and agent of economic vitality for the entire region. This can be easily observed via the arts, entertainment, and upscale shopping and restaurant linkages between San Francisco and peripheral cities such as Walnut Creek, Mill Valley and Palo Alto.

Cosmopolitan Culture

The cosmopolitan culture of the Bay Area is well known for its diverse cuisine, world-class wine, and great coffee; and most importantly its openness to new ideas, change, new people and a wide range of lifestyles. These are catalysts for a place that incubates creativity and innovation.

Bay Area cities such as San Francisco and Palo Alto provide another ingredient knowledge workers want: places for creative people to meet and interact face-to-face informally while enjoying their surroundings. These places provide them with a sense of community as well as opportunities for networking, which are critical ingredients for the success of an innovation economy.

Silicon Valley's arts and cultural scene has improved throughout the years. Between 1995 and 2004 the numbers of arts organizations in the Valley rose by 54%. In the midst and aftermath of the Dot-Com Bust between 2000-2004, art-oriented non-profit organizations increased faster in the Valley than in the state or the nation. This is a testament to the rising interest in arts and culture in Silicon Valley, despite the fact that it is a place more widely known for technological innovation.

Part 3

Silicon Valley's Sustainability

P erhaps the most unique feature of the Silicon Valley innovation economy is its resilience, despite continuous change that has included the high failure rate of start-ups, shakeouts such as the Dot-Com Bust and intense global competition. (Figure 22) Creighton University professor Ernie Goss, a scholar- in-residence at the Congressional Budget Office studying the economic impact of technology, says "Cutting edge tech hubs have to reinvent themselves every cycle. Reinvention and dynamism give them life and death. This is a natural cycle that can be brutally painful to workers and residents."[107]

Figure 22

Silicon Valley: Signs of Continued Health

- In 2006, 11 of the 20 most innovation-rich towns in the U.S. were in Silicon Valley

- Patents to Silicon Valley are rising:
    - Patents granted to Silicon Valley residents grew from 114 per 100,000 persons in 1994 to 377 in 2004 (a 330% increase)
    - Patents granted to the Valley increased 24% in 2006, the biggest single-year increase in a decade
    - Silicon Valley's number of patents per 100,000 persons is 14 times greater than that of the U.S.

- International co-inventing and collaboration in patents is also rising

Sources: Wall Street Journal, *Joint Venture: Silicon Valley Network*

An Environment of Continuous Change

Factors contributing to the region's longevity and vitality include the constant creation of new companies stemming from new ideas funded from diverse sources, including personal capital, angel investments and venture capital. Another key factor is the Valley's culture of the serial entrepreneur. "Silicon Valley recycles its talent... individuals embrace the start-up culture and become comfortable with the risk of failure," says Rolando Esteverena, a successful serial entrepreneur who came to the Valley from Argentina years ago.[108]

In her 1997 comparison of Silicon Valley and Boston's Route 128 area as innovation economies,[109] Virginia Postrel looked at resilience and adaptability as strategies adopted differently by competing regional economies. She found that these two regions have adapted different strategies in response to risk, change, and crisis.

Postrel concludes: "Boston has encouraged strategies of anticipation. People try to imagine everything that might go wrong and fix it in advance. But in Silicon Valley, there are no certainties. The future is open and subject to upheaval. Resilience is the strategy of choice. People do the best they can at the moment, deal with problems as they arise, and develop networks to help them out."[110]

Postrel uses the example of two early competitors in the workstation market: Sun Microsystems (a Silicon Valley-based company) and Apollo Computer (headquartered near Boston). At one point, Thomas Vanderslice, Apollo's chairman, characterized Sun's less-structured work atmosphere with a contemptuous cliché: "In this country, everything loose rolls to the West Coast."[111]

While Apollo Computer took a more cautious approach to entering the market selectively, moving forward only as much as it felt it could properly support (a reasonable business philosophy), Sun Microsystems took a much more aggressive approach by going after every opportunity available. When the semiconductor industry experienced a slowdown in 1984, Apollo found itself with a shrinking market while Sun had already developed a new, less expensive product line to attract new customers. Despite efforts to maintain its position in the market by cutting prices, Apollo declined and was eventually sold to Hewlett-Packard.

Apollo's methodical response to market changes proved too slow for the increasingly fast-paced marketplace it served, while Sun's more nimble strategy allowed it to adapt quickly in the fast-changing market conditions.

Waves of Innovation

And certainly, Silicon Valley has weathered storms. But with every "wave of innovation"[112] and economic boom and bust cycle, its economy has successfully managed to reinvent itself. (Figure 23) For much of the first half of the 20th century, Valley companies were deeply involved in developing and manufacturing defense-related products. In response to the recession in 1970, the Valley's innovators began applying defense technologies (particularly integrated circuits) to new commercial applications. A wave of electronic innovation swept through the region, laying the foundation of the personal computer that followed in the early 1980s.

Figure 23

Silicon Valley Waves of Innovation

Annual Non-Farm Employment, 1970-2007

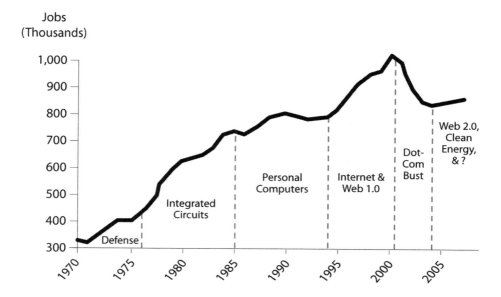

Source: Bureau of Labor Statistics, SiliconValleyOnline.org, Munroe Consulting, Inc.

With each cycle, disruptive technologies or depressed economic conditions (e.g., the defense recession of 1970 or the semiconductor recession of 1985[113]) changed the playing field and a new cycle of reinvention began (see Figure 23). Over the years, the Valley's economy showed tremendous resilience in rising from economic downturns and major business setbacks, best illustrated by the Dot-Com Boom/Bust cycle. On March 9, 2000, the NASDAQ stock exchange closed at an all-time high of over 5,000. The tech-heavy stock index had risen from 3,000 to 5,000 in four months.[114] (Figure 24) Many of these high-flying "new economy companies" were based in Silicon Valley or San Francisco.

But there were serious problems. The prices of a large number of Internet-based stocks could not be justified on the basis of their financial performance or the soundness of their business plans. The chasm between valuation and performance led to the collapse of Internet stocks. In March 2000, two trillion dollars worth of equity wealth vanished in a matter of a week; hundreds of e-businesses failed, and hundreds of thousands of jobs were lost in the process.[115]

On the tail of the 2000 Dot-Com Bust was the collapse of the tele-communications market in 2001 as a result of the industry's over-investment in infrastructure that greatly exceeded actual versus anticipated demand. The telecom meltdown contributed to the Valley's economic collapse and hindered a quick recovery.

While the Dot-Com Boom made Silicon Valley a household term, the Dot-Com Bust of 2000-01 showed the world that the Valley is not immune to market forces and that economic prosperity can be fleeting. Underlying one of the greatest financial disasters of our time was a euphoric view of the future with enormous expectations involving the Internet. The financial markets, at least for a while, supported the spectacular promise of the Internet. Investors paid astronomical prices for shares in Internet-related companies.

After the sharp, hockey stick-shaped rise in 1999 and 2000 stemming from the Dot-Com Boom, VC investments dropped equally as fast because of the bust in 2001, 2002 and 2003. Venture capital firms licked their wounds and waited

Figure 24

Dot-Com Boom/Bust Cycle
NASDAQ Index, 1990-2007

Source: NASDAQ

to see where the markets were going next. As dramatic as the Dot-Com Bust was, VC investing in Silicon Valley (in dollars) never dropped below the Q1-1998 level. Between 2003 and 2007, there was a steady increase in the total amount of venture capital money invested in Silicon Valley companies and the total number of deals.[116] (Figure 24) Media pundits were heralding, "Silicon Valley is back!"

Resurgence: The Next Big Thing

The years following the bust have demonstrated the adaptability and sustainability of the Valley's innovation economy in the face of economic adversity.

In a matter of a few years, venture capital investment levels climbed back to pre-bust levels. The Internet did not "collapse" and disappear by any means; in fact, it spawned viable Internet-related businesses based on the lessons learned from the Dot-Com disaster. Now Web 2.0 has taken off and productivity levels continue to climb, along with the adoption of new technology.[117]

Silicon Valley has also stepped up to meet the growing challenge of global warming and the rising cost of fossil fuel. The Valley has the potential to become the leading region for clean energy products. That is so thanks to its highly sophisticated innovation ecosystem, including the region's access to major research universities, think tanks and entrepreneurs; its critical mass of highly successful technology companies; and its highly sophisticated venture and angel capital presence.

The Web 2.0 Surge

Web 1.0 was the birth of our love affair with the Internet. It featured websites, browsers and browser wars, e-commerce and instant access to all the data you'd ever want or need (or so you thought). It spanned the Dot-Com Boom and the Dot-Com Bust through the 1990s. But it is now history.

The economy's recovery in 2004 and 2005 gave VCs time to recover and reconsider their next moves. Given the new breed of Internet entrepreneurs with more robust business plans and given the fact that most of the world's Internet users are now thoroughly hooked on the Web, we have the birth of Web 2.0.

What is Web 2.0? It is, as they say, you. It features user-driven content and interaction. It offers multimedia, homemade videos, blogs, social networks and, most important for Web 2.0 entrepreneurs, it delivers advertising. Early on in the life of the Web, many new Silicon Valley start-ups stacked their bets on e-commerce as the ticket to fame and fortune, selling everything including pet food (Pets.com), and groceries (Webvan.com) online. Unfortunately, sales didn't always support the required huge investments in infrastructure, software development, marketing and highly-paid engineers. Web advertising was certainly part of the plan, but the model was flawed: Hit rates were sky-high, but clicks didn't always translate to sales at the rate expected. And Web 1.0 advertising was often circular—one dot-com company would sign an agreement to advertise with another and vice versa. Both would report the deals as revenue—not a sustainable model.

Today's Web 2.0 businesses are taking a much more practical approach. Now that almost 50% of U.S. households are connected to the Internet via high-speed broadband cable or DSL services, Web 2.0 companies can offer a much richer online experience by offering music, animated images and streaming videos. A couple of years ago, accessing such high-bandwidth websites may have been possible via dial-up service, but at 56K Mbps, the user experience would have been far less satisfactory.

Music has been one of the leaders of the Web 2.0 surge. Oakland-based Pandora is the third-largest Internet radio station in the world, with 6.9 million registered users who have played over 4.7 billion songs since the company's founding in 2000. One of the keys to Pandora's success is the company's team of fifty music analysts who index every song by over 400 attributes, from melody, harmony and rhythm to instrumentation, orchestration, arrangement, lyrics, and of course the rich world of singing and vocal harmony. As users listen to their favorite songs, Pandora's automated music-matching system recommends complementary songs that the listener may like. The company generates revenue from the sale of songs online as well as from online advertising and product sales. Pandora is also positioning itself at the center of the convergence of hand-held devices by offering its music streaming service via Pandora-enabled cell phones on a monthly subscription basis.

Another example of the changing ways we interact with the Web is the social networking phenomenon. While the pre-teen- and teen-oriented MySpace.com has been a media star for the Web 2.0 revolution, Palo Alto-based Facebook.com early on established itself as a popular social networking site for college students. It has since become widely used by business professionals. Facebook currently has over 175 million active users. Demonstrating the kind of creative business agility needed to survive in the world of Web 2.0, Facebook has added a variety of new features for sharing photos, posting blogs, downloading music and posting free classified ads in direct competition with the widely popular Craigslist.org. Facebook's internal valuation is estimated to be around $8 billion with projected revenues of $1 billion by 2015.

A third example of how user-posted media content is driving Web 2.0 is Flickr.com, a very popular online photo-sharing Web site. Launched in 2004 and formerly based in Vancouver, BC, Flickr is now headquartered in the Silicon Valley city of Sunnyvale. Yahoo! acquired Flickr in March 2005 after it saw the business as a much more advanced Web 2.0 service than its Yahoo! Photos service. One of Flickr's revenue-generating strategies is to let users post and sell their photos online. This opens up tremendous opportunities for amateur and professional photographers to monetize their pictures, while effectively converting Flickr's millions of registered picture-takers into content producers.

Flickr was the first photo-sharing site to allow users and viewers to tag images with keywords to facilitate searching and sorting. This type of use of user-generated metadata for defining and categorizing images is often cited as an early example of "folksonomy"—a term (derived from *folk* + *taxonomy*) referring to the practice and method of collaboratively annotating and categorizing web content based on creator and consumer keywords, rather than simply a pre-defined list of subject indices.

Web 2.0 has successfully reawakened VC interest in Internet investing and is going global. In the second half of 2007, investors closed 101 Web 2.0 deals worldwide worth over $464 million—a 7% increase in dollars over the same period in 2006 and a 14% increase in the number of deals.[118] While the San Francisco Bay Area garnered the largest number of VC investments (25 deals, $91 million), other U.S. regions (most notably New England and New York) are beginning to challenge the Bay Area's leading position. VCs closed ten deals in the New England area worth over $102 million.

Other nations are also becoming significant competitors to Silicon Valley in the Web 2.0 arena. Israeli companies took in $15 million in five deals, while the U.K. received $22 million in seven deals. China posted nine Web 2.0 deals ($41 million).[119] (Figure 25)

Figure 25

VC Investments in Web 2.0

Worldwide, July–December 2007

Region	Deals	Investments ($ millions)
Bay Area	25	$91
Europe	20	$52
New England	10	$102
China	9	$41
Southern California	8	$59
U.K.	7	$22
Israel	5	$15
Other	17	$82
Total	101	$464

Source: Ernst & Young LLP, September 2007

The Cleantech Revolution

The critical importance of the emerging cleantech industry cluster cannot be overstated. We have enormous global challenges such as climate change, energy dependence on volatile regions of the world, and health and economic damages resulting from our fossil fuel-burning economies. Silicon Valley's keen interest in cleantech comes not only in response to serious challenges to our environment, our health, our lifestyle, and our need for energy independence; but also in a drive to establish its leadership in a vast array of businesses that can be very profitable.

So the Silicon Sunrise is on.[120] The financial markets have continued to endorse the business case for solar power for the past several years. In 2005, three of the biggest initial public offerings involved three solar power companies, including SunPower, a spin-off company of semiconductor manufacturer

Cypress Semiconductors. SunPower raised $146 million when it went public in 2006. Sales growth has been impressive: 2006 revenues are expected to reach $230 million, a huge leap from 2004 revenues of $10 million. The venture capital industry has seen the light, and that is true nationwide: VCs invested more than $1.4 billion in solar power in the first six months of 2006 compared with nearly $1.4 billion in all of 2005.[121]

Despite intense global competition, Silicon Valley has continued to stay ahead of this game as well. The region is once again becoming a global leader, this time in cleantech,[122] aided in part by California's strong pro-environmental policies.[123] California's statewide "green" initiative requires the state's two huge public pension funds to commit $1.5 billion to environmentally-friendly investments.[124] Silicon Valley is increasingly focused on advancing clean energy generation, including solar power, biofuels, fuel cells and hydrogen power. It is estimated that nearly 30% of all VC investments in the clean energy sector are made in Silicon Valley companies. The region resonates most with solar technology.

There are several reasons why:

- Technological familiarity. Semiconductor and solar industries share the same silicon-based technology.
- Improvements in silicon technology, falling costs and the promise of cost parity with fossil fuel or nuclear-powered electricity generation.
- Strong interest in decentralized electricity generation.
- Enormous solar energy business potential. The worldwide market is expected to rise from $11 billion in 2005 to $51 billion in 2015, according to Clean Edge, Inc.
- Renewable energy business potential, expected to be over $167 billion by 2015.
- Favorable legislation: California law calls for the installation of 1 million rooftop solar panels (worth $2.9 billion) on businesses, homes, schools, farms and public buildings between 2011 and 2018.
- Solar power comes from a free and inexhaustible source and works even on cloudy days.

In 2005, Silicon Valley companies closed 23% of the venture capital deals for clean technology development and 5% of all cleantech deals in the United States. VC investments in the U.S. clean energy sector rose from $552 million in 2004 to $884 million in 2006—a 60% increase in two years. Investments in Silicon Valley clean energy companies also accelerated. Between the first quarter and the third quarter of 2006, VC investments in clean energy increased from $50 million to $300 million (600% growth in six months), indicating strong interest in the potential of this emerging industry. As a region, the Valley outpaced other tech regions of the United States by receiving a third of the total VC investments during that period.[125]

In 2006, Silicon Valley alone closed more cleantech deals than the state of Maryland, which ranks third among states in the U.S. for cleantech investment after California and Massachusetts.[126] The Silicon Valley venture capital firm Kleiner, Perkins, Caufield & Byers planned to spend $600 million of its investment fund in 2006 on projects that reduce CO_2 emissions. The firm expects to invest nearly a third of all its new funding in cleantech by 2009. By the end of 2007, venture capital investments in Silicon Valley cleantech projects totaled $1.1 billion.[127] (Figure 26)

In the third quarter of 2007, Silicon Valley companies received an impressive $2.48 billion in VC investments. Some of the major deals in the third quarter of 2007 were in cleantech. The biggest deal in 2007 was the $200 million obtained by Shai Agassi, an ex-SAP executive and rising Valley entrepreneur, for his start-up business. His firm, Better Place, focuses on 21st century electric car designs.

The sun is back in favor again as a source of electricity generation. Several factors are in play, and they include volatile oil prices, energy independence, heightened concerns about greenhouse gas emissions and a requirement in at least 25 states, including California, that utilities buy power from renewable sources and develop new solar technologies.

The solar energy market today is dominated by photovoltaic, or PV, solar power that directly converts the sun's rays into electricity via silicon panels. The PV solar power business is booming. One indicator of its popularity is that major

Figure 26

VC Investments in Silicon Valley Cleantech Companies
2005-2007

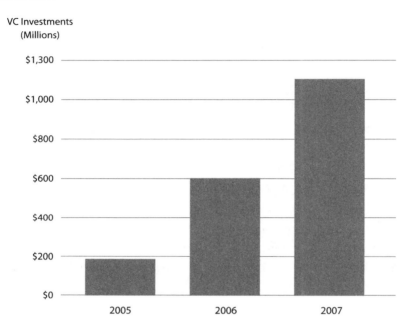

Source: Cleantech Group, LLC

American corporations such as Wal-Mart, Microsoft, Google, Estée Lauder and Tiffany's are planning to install solar panels on their rooftops to supply electricity to their businesses.

An alternative form of solar power, solar thermal, has enormous potential. It has been around for a long time, just as solar PV has. The promise of solar thermal is that it can be used to generate electricity in large quantities. Hence, it can be used to run electric utility power plants.

A solar thermal system consists of a large array of mirrors, which focus the sun's rays to heat water or another fluid. This generates steam, which drives turbines, similar to a conventional power plant, and generates electricity. As a source of large-scale power generation, thermal solar is less expensive and more practical than PV solar. It is important to remember that although both types of solar power systems produce electricity, solar PV produces electricity directly. A thermal solar system first produces heat and then electricity.

One frequently asked question is what happens to power generation from a solar thermal system when the sun does not shine. The answer is fairly simple. As the thermal solar plant first generates heat, that heat can be stored before it is used for electricity generation. When the sun goes down, the stored heat can be used to turn the turbines that generate electricity.

When a thermal solar plant is located at a place with an abundance of sunshine, the plant becomes very reliable. Much of the American West is an ideal location for thermal solar power plants.

A cluster of large solar thermal plants based on parabolic trough technology was built by Luz Industries in the 1980s in the Mojave Desert. The solar farm covers four square miles, has 400,000 mirrors and annually produces 354 megawatts of electricity, sufficient to serve tens of thousands of homes. Not much happened in the two decades since that plant was built until natural gas prices escalated rapidly.

In June 2007 in Boulder City, Nevada, 35 miles southwest of Las Vegas, a new solar thermal power plant called Solar One went online at a 280-acre desert site with 220,000 giant glass mirrors faithfully tracking the sun's progress across the sky. It is a sight to behold in the bleak Nevada desert. The large Spanish conglomerate Acciona built the plant, and it produces 64 megawatts of electricity a year.

In July 2007, Pacific Gas & Electric entered into a major renewable energy agreement with Solel Solar Systems of Israel to purchase energy from the Mojave Solar Park, to be constructed in the Mojave Desert. The Solel Solar plant will be one of the largest in the world. PG&E has contracted to obtain 553 megawatts of solar power, sufficient to serve 400,000 homes per year.

Solel Solar Systems is the prime thermal solar company in the world. Solel's patented solar thermal parabolic trough technology powered the first solar plants in the Mojave Desert. When fully operational in 2011, Solel's new plant will cover 6,000 acres (nine square miles) in the Mojave Desert.

In October 2007, PG&E signed a contract with Ausra, Inc. of Palo Alto to build a large solar thermal plant in California. More details on Ausra are provided in the section below.

The good news is that nonpolluting solar thermal power is back, and utilities can build power plants to generate it instead of consuming coal or natural gas.

The enormous potential for solar thermal in the United States will be fully realized when it can generate electricity at 10 cents a kilowatt-hour or less, the current average price of electricity in the United States. That day may not be too far off.[128]

Silicon Valley's contribution to the clean energy challenge may prove to have far greater impact than even the PC or the Internet. Several Valley solar energy companies stand out in terms of making a real difference:

SolFocus

Launched in 2005, SolFocus is dedicated to developing solar modules that will lower the cost of solar arrays—a critical need if solar power is to succeed commercially. The first goal is to use less silicon, which is expensive because of supply constraints (more silicon is now used in the making of solar cells than for computer chips). The second goal is to make the cells more efficient. Both of these are accomplished via the design of the solar modules that incorporate a system of curved mirrors and lenses that magnifies sunlight onto solar cells made of germanium (another semiconductor with properties similar to silicon). There are other Valley companies such as Solaria and Pacific SolarTech that have objectives similar to SolFocus: to make solar power competitive with electricity generated from fossil fuels.

Nanosolar

The poster child of Silicon Valley solar power companies, Nanosolar was founded in 2001 and attracted super-angel investors such as Google founders Larry Page and Sergei Brin and the Austrian Internet entrepreneur Martin Roscheisen. It was also the first solar power company to raise Sand Hill Road venture capital money. The company is well-funded and so far has raised more than $100 million. It is developing a technology that involves fabrication of solar cells made from thin films consisting of copper, indium, gallium and selenium (CIGS).

Applied Materials

One of the most well-known Valley chip-making equipment companies, Applied Materials has announced plans to sell tools and machines for fabricating solar cells, as it expects the market for such machines and tools to top $3 billion in the next four years. The company also plans to use its technology for making flat-panel displays for the production of thin-film solar cells sprayed on glass and other surfaces. The overall objective is to make solar power more competitive by lowering the cost of fabrication.

Solyndra

Founded in 2005 and based in Fremont, California, Solyndra has raised more than $600 million in venture capital financing. In October 2008, the company had nearly 600 employees. The firm started shipping their products in July 2008 and now has orders for $1.2 billion worth of solar panel orders to be delivered over the next five years. The company entered the market with an innovative design: a cylindrical photovoltaic solar cell made with thin films of copper, indium, and gallium arsenide to create semiconductor material. These cells are ideal for the flat rooftops of commercial buildings around the world. According to company officials, if the commercial rooftops in the U.S. were covered with their solar cells, they would generate enough electricity to power nearly 15% of the homes in the U.S. The company's PV cell design cuts down installation costs by nearly 50% and installation time by 75%. This is a key advantage, as installation costs are a major factor in solar PV systems.[129]

Ausra

Ausra designs, manufactures, and markets solar collectors and steam generation systems. Additionally, the company provides solar energy generation and distribution services. The company uses relatively inexpensive 40-foot long flat mirrors that concentrate solar rays on water pipes to produce steam for generating electricity.[130] Ausra received significant funding from the major Silicon Valley VC firms of Khosla Ventures and Kleiner, Perkins, Caufield & Buyers. The company has a 5-megawatt plant in Bakersfield, California, and generates enough electricity to power 3,500 homes in central California. Several other solar plants are under construction.[131]

Perhaps the most startling and exciting development was the November 2007 announcement that the iconic Silicon Valley company Google was stepping into a very different business from what made it famous: the Internet search business. Google plans to spend tens of millions of dollars over the next several years on research and development relating to advanced solar thermal power, wind power, geothermal systems and other state-of-the-art energy technologies. This diversification move makes good business sense for Google in light of its enormous capacity to innovate by hiring some of the best scientists and engineers in the alternate energy field.

Clearly, Silicon Valley has successfully maintained its economic viability over many decades despite continuous change, including the high failure rates of start-ups, economic shakeouts such as the Dot-Com Bust, and intense global competition. The region has demonstrated its resilience through the constant creation of new ideas and new inventions. The region's innovation economy has rebounded time and time again through the efforts of serial entrepreneurs who have formed new companies funded from diverse sources to bring innovative products and services to market. In the face of constant change and increasing competition, Silicon Valley has embraced the start-up culture and become comfortable with risk as well as reward. The Valley's undaunted spirit is a foundation of its continued success.

Innovation Excellence Continues

Another vital sign of the Valley's resurgence as a center for innovation excellence and a technology hot-spot is the fact that several new companies received nearly a third of the World Economic Forum's prestigious Annual Technology Pioneer Awards for 2008.[132] The U.S. as a whole captured 23 of the 39 awards. The other countries sharing the remaining 16 awards included Israel, Great Britain, Sweden, Switzerland, Canada, France, Germany, India, the Netherlands, and Russia. Out of the 39 total awards, Silicon Valley companies obtained 11. Winning companies must demonstrate that their innovation has the potential for evolving into a life-changing technology as well as making them a market leader.

The Silicon Valley winners and their key products were:[133]

+ **23andMe** (Mountain View) Offers genetic testing and personal DNA analysis that allows people to search their genetic makeup.
+ **Acuray** (Sunnyvale) Produces a doctor-directed robot called CyberKnife that treats tumors with radiation. To date, nearly 40,000 patients have been treated worldwide.
+ **Admob** (San Mateo) Connects advertisers with mobile phone publishers to allow advertisers to create and target ads effectively.
+ **Arteris** (San Jose and Paris) Develops networking capabilities for chips that have multiple functions built into them.
+ **Innovative Silicon** (Santa Clara) Develops and licenses low-cost memory technology, now seen as a successor to DRAM (dynamic random access memory).

In a July 2008 conversation, Peter Darbee, CEO and Chairman of the Pacific Gas & Electric Company (a leader in the development of solar thermal power in California) said, "Among the many promising renewable technologies we are exploring, solar thermal is one of the most attractive, with capabilities that match up well with California's energy needs. We see it as a vital contributor to meeting the state's clean energy and climate change goals."

Tapan Munroe

Source: "Oil Price Puts Solar in the Spotlight," Contra Costa Times, *July 20, 2008*

- **LS9** (San Carlos) Re-engineers microbes to act like those in petroleum and produce renewable petroleum—the ultimate goal is to produce bio-fuels.
- **Lumio** (Menlo Park) Produces low-cost sensing technologies that can convert an inert surface into an interactive touch-sensitive area.
- **Meraki** (Mountain View) Provides affordable Internet access via a new approach to wireless networking.
- **Nanostellar** (Redwood City) Applies nanotechnology to catalytic converters to help clean diesel engine exhaust.
- **Silver Springs Networks** (Redwood City) Produces intelligent, protocol-based networking products.
- **Unidym** (Menlo Park) Makes products for the electronics industry using carbon nanotechnology.

Challenges and Threats

Silicon Valley faces many challenges that could undermine the health of its innovation ecosystem. These challenges include intense global market competition, competition for world-class talent and investment capital, declining financial support for the great public research universities such as UC Berkeley and UC San Francisco, and a decline in the region's quality of life.

However, we feel that the greatest challenges surround the declining quality of high school education in the region, the shortage of mathematics and science graduate students for the region's universities, worker burnout, local traffic congestion and high housing costs. Not paying attention to these challenges and failing to find solutions will certainly pose a threat to the region's economic health. (Figure 27)

Figure 27

Risks to the Silicon Valley Ecosystem

- Loss of talent
- Declining quality of U.S. secondary education
- High housing costs
- Declining quality of life
- Transportation congestion
- Worker burn-out
- Domestic and international competition
  — Increasing VC investments in Boston area Cleantech companies
  — Increasing innovation around the globe, e.g., India, China

Source: Munroe Consulting, Inc.

Lack of High School Performance

Although Silicon Valley companies such as Google, Yahoo!, Cisco Systems, HP and others attract the best talent from all over the world, and the region is endowed with some of the finest research universities in the world, the Valley continues to be challenged by the lack of performance of the region's high schools.

This is a critical concern for such a knowledge-intensive region. The region's ability to prepare its high school graduates for entrance to the various campuses of the UC system, Stanford University, the California State University (CSU) campuses and other universities in the nation ties directly to the future economic success of the Valley as well as the San Francisco Bay Area. The region's companies must be able to hire top talent locally and regionally, not just globally. Data from 2004 to 2006 suggests that Silicon Valley high school graduation rates have dropped over the last two years. A matter of greater concern is the fall in the portion of graduates who did not meet the requirements for entry into the UC and the CSU systems. This is a reversal from previous years. In addition, the Valley's high school dropout rates have continued to increase for the last several years.

Shortage of Mathematics and Science Graduates

In 2006, an estimated $2.6 billion in venture capital poured into start-ups, including software, clean energy, biotechnology, and Web 2.0 businesses. High-quality science and engineering graduates are critically needed by these fledgling start-ups, whose survival relies on creating ideas and innovations and shaping them into new products and services. These grads are just as valuable as venture capitalists and entrepreneurs.

However, there is a major concern about the sustainability of the region's innovation ecosystem due to the local talent crunch. There is fierce competition in Silicon Valley for hiring and retaining the best talent. Tech companies such as Google, eBay and Yahoo! are searching not only globally but also locally for the best and brightest talent. It is the local picture that is of the greatest concern.

The real problem for California, the Bay Area, and Silicon Valley is that not enough engineers and scientists are graduating from local universities to meet the demand of high-tech industries. The source of this problem is that a small and declining percentage of high school graduates are entering technical fields that require science and mathematics.

This problem is much more serious than students' lack of interest in these subjects. There are several reasons why most students are not taking science and math courses today. These include:

- **Poor performance** A National Science Foundation report indicates that California's eighth-grade students placed last in science and near the bottom in math compared with students in other states.
- **Disinterest in hard work** Many students want to coast through easy courses. At the same time, they want to become hedge fund managers or investment bankers and make lots of money fast, a highly unrealistic view of the world.
- **Lack of technical jobs** This is an erroneous perception. On any given day, Bay Area job listings on Craigslist.org suggest there are several thousand technology-related job vacancies in the region. These jobs typically pay very well. The average tech salary in Silicon Valley is more than $90,000 a year.
- **Fear of being bored** Students who have grown up on MySpace, YouTube, iPod and television often say they find math and science dull in comparison.
- **Teacher shortage** In 2002-03, an estimated 1,500 mathematics and science teachers with no credentials taught more than 800 science classes in California high schools. This is a serious problem.

Another serious concern arising from the lack of students entering university science and math programs is that many of these programs may be closed, due to lack of enrollment. The loss of innovative science and math programs limits opportunities for truly talented students and continues the downward spiral of educational achievement.

Various programs being instituted may alleviate this problem down the road, but reversing these trends will take time. Without a doubt, the most ambitious teacher development program in the nation is the California Teach program, unveiled in May 2005 by Governor Arnold Schwarzenegger and UC officials. The program's goal is to quadruple the state's annual production of science and math teachers from 250 to 1,000 by 2010, and thus far it appears to be on track. This initiative will encourage UC students interested in science, mathematics or engineering to consider teaching as a career. Essentially, the program is a public-private partnership among the state, UC, California State University, elementary and high schools and 18 large corporations including Adobe, Amgen, Boeing Chiron, Intel, Qualcomm, Sun Microsystems and U.S. Bank.

Despite challenges that include program funding and the retention of good science and math teachers, we are optimistic about the program's long-term success. However, it will take years before California and its technology regions such as Silicon Valley rebuild their corps of science and math teachers, and more importantly, produce sufficient numbers of science and engineering graduates to meet the hiring needs of Silicon Valley and other technology regions of the state.

Worker Burn-out

Despite Silicon Valley's high wages, high quality of life, and supportive social and professional networks, its fast pace and competitive nature makes it not only a leading technology region but also a leader in worker stress. Because of the intense pressure to stay on the cutting edge of technology and business, long workdays have become the rule rather than the exception. At the peak of the Dot-Com Boom in 2000, Stephanie Brown, director of the Addiction Institute in Menlo Park, noted marked increases in stress-induced psychological problems among information technology and other professionals in Silicon Valley. The problems included marital strife, alcoholism, and an addiction to work.[134]

The Valley has a dominant "culture of accomplishments," where being first in the world is important. According to David Gamow, Founder of Clarity Seminars in Mountain View, the Valley moves quickly, and if everyone is moving at 200 miles an hour, then 100 miles an hour is slow. *More speed* is the mantra of the Valley.[135]

In high-stress places such as Silicon Valley, the factors that contribute to overwork include:

+ Lack of control over workload or speed of work
+ The global 24/7 economy, enhanced by work shared by workers located all around the world
+ Job insecurity, stemming from volatility in the economy and memories of the Dot-Com Bust
+ Less time for friends and family
+ Less emphasis on health and exercise

Overwork and burnout is not just a Silicon Valley phenomenon, it's an American one. According to research from the Sloan Work and Family Research Network at Boston College:

+ One in three American employees is chronically overworked (2005)
+ Stress-related illnesses cost American employers about $300 billion a year[136]
+ Fewer than 10% of organizations say they are doing well with managing workload (2004)
+ In the last 25 years, the combined weekly hours of all couples has increased from 70 hours to 82 hours, while for dual working couples with children under 18, combined hours have gone up from 81 to 91.[137]

And with the current economic crisis, it is evident that workers face further pressures as they are asked to do more with less and they cope with fewer workers and increasingly limited resources. This is what usually happens during a recession. However, this pressure also results in higher productivity, and eventually sets the stage for an economic recovery.

Joe Robinson, author of the book *Work to Live*, notes that Americans are the hardest-working people in the world and work the longest hours in a year: 350 hours (or 9 weeks) more than workers in western Europe. Only 14% of Americans get two weeks of paid vacation per year. By contrast, workers in Austria and Finland get six weeks of paid vacation, and workers in Germany,

France, Spain, and Italy get five weeks of paid vacation a year.[138] Robinson was a participant in a panel discussion in Santa Clara sponsored by an organization called Take Back Your Time. He said that he was there to raise awareness of the overwork and burnout problem because the Valley is the "capital of overwork and the 24/7 economy."[139]

Granted, not everyone feels overworked in the Valley. Many people enjoy the long hours as they have learned how to lead a balanced life. Personal perceptions of workload vary depending on individual needs, lifestyle, expectation, and experience. Also, individuals have differing tolerances for demand and stress. However, on the average, large numbers of workers in the Valley are overworked. This results in low employee retention rates, lower efficiency, and loss of creativity. These are serious problems for idea- and creativity-intensive companies, and a potential threat to the sustainability of the Valley's innovation ecosystem.

Traffic Congestion and High Housing Costs

Rapid economic growth has resulted in increasingly clogged freeways and skyrocketing housing prices in Silicon Valley. Traffic congestion and housing affordability impact the region's ability to attract top talents. As we have noted, this is a serious concern expressed by 115 Silicon Valley CEOs, according to a recent business climate survey conducted in May 2007 by the Silicon Valley Leadership Group, an organization with a membership of 210 Valley companies.[140] Nearly 84% of the respondents said that the high cost of housing and traffic congestion were major concerns relating to the cost of doing business in the Valley. With the median cost of the Valley housing at $800,000, only 15% of the region's families could afford a home in Silicon Valley. Some of the responding CEOs suggested that the Valley's high cost of housing is turning out to be a deal-breaker in the process of hiring new talent.[141]

Growth has also meant more cars, more congestion, and longer commutes to this vital region. Traffic congestion means commuters are wasting their time sitting in cars, smelling polluted air, and worsening the air quality of the region. In 2005, rush-hour traffic delays cost commuters $973 and wasted 54 hours per capita in the San Jose area. The cost was above the national average of $710 per commuter. According to the Texas Transportation Institute, the total economic cost of congestion to the San Jose economy that year was $899 million.[142]

It is important to note that despite its twin problems of high home prices and traffic congestion, the Valley remains the top global technology hub. It dominates technology and life sciences industries over other global technology hubs such as Singapore, Shanghai, Tokyo, Basel, Bangalore, Dublin, and Berlin. Why? In part because in many of these places, the cost of housing and traffic congestion is even worse than it is in Silicon Valley.

A Culture of Quick Fixes

According to Judith Estrin, a former Cisco Systems executive, the Valley has increasingly become focused on quick fixes and quick returns.[143] This has weakened a fundamental process vital to the sustainability of the Valley's innovation ecology: the process of transforming basic research into viable products and services that add value to our lives. Estrin notes that there is less emphasis being placed on basic research and companies are spending more time on longer-term development of products. Forces such as Wall Street's emphasis on corporate quarterly financial statements and declining public sector investment in research and development have weakened the source of transformational developments, ranging from the invention of the transistor to the development of the Internet. She suggests that corporate leaders used to be more interested in building companies and creating jobs and not just starting and flipping companies to make a fast buck. This is undoubtedly a threat to Silicon Valley's innovation ecosystem.

Similar concerns about the future of the Valley have been expressed by Valley leaders like retired Intel CEO Andy Grove. Grove is very concerned about what he calls the "exit strategy" problem. This has to do with the business philosophy of leaders of start-up companies who tend to be fixated upon selling a company to the highest bidder instead of building the company over the longer term.[144] Corporate leaders used to be more interested in building companies and creating jobs and not just starting companies with the goal of selling them off quickly. The fast buck trend started with the Dot-Com Boom in 1998. Although the Valley recovered from the Dot-Com Bust that followed, the culture of quick returns still persists to a considerable degree. This is not how the anchor companies of Silicon Valley—such as Hewlett Packard, Intel, Google, Cisco Systems, Oracle or Apple—were built.

An Ecosystem in Flux

The good news is that innovation is alive and well in the Valley. But a trend is emerging: in 2007 and 2008, most of the trail-blazing innovations were created by large companies. Based on the economic realities of Silicon Valley in 2008, the questions we raise are these:

- Can the Valley's innovation ecosystem be sustained when only larger companies can afford to engage in innovation activities? Where will funding for small start-up companies come from, if VCs continue to shift focus to less risky later-stage companies while the cost of R&D continues to rise?
- The magic of the Valley has always has been found in the creative start-ups (such as those profiled above) that ultimately become global leaders. What will happen to the economic health and entrepreneurial culture of the venture capital industry if VCs are not engaged in funding highly innovative start-ups?
- Does this emerging trend support or undercut Silicon Valley's stellar record of innovation?

Our conclusion at this point is that the sources of innovation in the Valley are changing—the Valley's structure of innovation is evolving to maintain the health of its ecosystem. We are seeing more innovations from public-private partnerships as well as greater contributions by larger companies. As we have noted, the promising areas of innovation in the Valley are likely to be in cleantech and health-related fields. These are most likely to be the sources of the next big things from the Valley in the first 25 years in the 21st century.

Surviving Economic Storms

A question arises each time the Valley goes through an economic downturn (e.g., recessions) or serious shifts in market conditions (e.g., declines in defense spending, the collapse of online revenues). Will the Valley's innovation ecosystem survive?

As we have seen numerous times, the Valley's economy is quite resilient and able to rebound from sudden economic shocks and rapid changes in its business environment. Historically, the Valley has weathered external shocks well because of its highly trained and educated workforce. It also managed to survive a serious self-inflicted wound, the Dot-Com Bust, through what one could call a complete system reboot.

The global financial crisis of 2008-2009 is unprecedented in the economic history of Silicon Valley. The U.S. economy went into recession starting in December 2007 as a result of the meltdown of the housing and the financial sectors. Most analysts expect that a recovery is unlikely to begin before the end of 2009. The threat to the Valley's viability is more serious than ever before. So once again, we must raise the question: Is the Valley sufficiently insulated from the economic and financial storm raging in the U.S. and the rest of the world to maintain its vitality and viability? How can the Valley weather the intensity of this economic hurricane?

Surviving the 2008-09 Economic Hurricane

Undoubtedly business leaders in the Valley are concerned about the impact of the global economic meltdown on the region, despite its record of resiliency. The region, after all, is deeply interconnected with the U.S. and global economies. It would be unrealistic to expect that the Valley will come through the current economic storm unscathed. It is not a matter of whether there will be an impact, but how much.

A large number of start-ups in the Valley are in the process of downshifting from overdrive to low gear in order to cope with the current economic crisis. The legendary emphasis on speed in capturing market share during the Dot-Com euphoria days has been replaced by caution and a focus on survival by management. Khris Loux, the founder of a software start-up known as JS-Kit, recently said, "You have to divide your loaf of bread and jug of water into small rations to get you as far as possible."[145]

A 2008 report from the Connecticut-based research firm Gartner, Inc. suggests that declining worldwide sales of computers and other gadgets, ranging from iPods to large servers, have led to decreasing demand for computer chips and caused economic problems for the Valley. This is because 20% of the revenue generated by the 150 biggest Valley firms comes from the sales of computer chips and chip-making equipment. The Gartner report expects that fourth-quarter 2008 chip sales will be down by 24% relative to the same period the previous year. The outlook for chip sales is not expected to improve until well into 2010.[146]

It is no surprise that layoffs and cost-cutting efforts have been announced by many major Valley corporations including Cisco Systems, Hewlett Packard, Google, Intel, Applied Materials, AMD, Microsoft, Intuit, Sun Microsystems, Symantec, Yahoo!, Agilent Technologies, National Semiconductors, and SAP.[147] The Internet auction giant, eBay, reduced its workforce in October 2008 by 10% when it terminated a thousand workers. eBay is also planning to eliminate several hundred temporary positions in order to cut costs and streamline operations. eBay attributed these layoffs to slow growth in its core auctions business.[148]

Strategies for Survival

Belt-tightening, layoffs, and hiring freezes at many Valley companies will continue into 2009 as the nation's economy remains mired in a recession. But while Silicon Valley is not immune from recessions, the Valley will fare better than most regions in the U.S. because of its innovation- and idea-based economy.

It is also likely to recover faster than most regions in the U.S. The Valley's resurgence at the end of the current global economic and financial crisis will likely be ushered in by cleantech and life sciences industries. Let us review a few of the factors that support the Valley's recovery and survival.

Factor 1: Start-Ups Are Smarter

The real concern in the current economic downturn is not about giants such as Intel, Google, and the Hewlett Packard's of the region. The concern is about the survival of the Valley's start-ups.

Job growth and sustainability in the region stem from the start-ups—they are the core strength of the region's economy. The hard-earned lessons of the Dot-Com Bust have not been lost and have been taken to heart by many Valley entrepreneurs. Scott Duke Harris interviewed nearly two dozen company founders of start-ups and summarized their comments.

His findings[149] suggest that:

- Prudent firms will survive the economic crisis crisis, even though they will no doubt feel its impact
- A recession is a good time to start a business (let us not forget that Google really began to climb its remarkable growth curve in the aftermath of the Dot-Com Bust)
- *Cut costs* and *small is beautiful* are the current mantras of business executives

Bright Spot: The Global Reach of Corporate Venture Capital

Intel Capital, started in 1991, is one of the largest corporate venture arms in the world. By the end of 2008 the venture fund had invested $7.5 billion in more than 1,000 firms.

Some of the success stories include CNet Networks (sold to CBS), MySQL (sold to Sun Microsystems), Red Hat, and VMware. In 2008, the firm invested $1 billion in Clearware, a new company that is planning to build a national mobile wireless network based on Intel's WiMax network protocol.

Not surprisingly, the strategy is part of Intel's objective of worldwide use of the company's products. It is impressive to note that Intel Capital so far has invested in 45 countries and maintains nearly 50% of its staff outside the U.S.

The CEO of the company, Arvind Sodhani, recently wrote to its portfolio companies, "Innovation does not stop during slowdowns. Nor should you! Innovation attracts customers… We are not wavering."

Deborah Gage

Source: Deborah Gage, San Francisco Chronicle, November 23, 2008

- Strategically savvy executives are focused on expanding their customer base and profitability, not on a mad rush to an initial public offering (IPO)

These perspectives make a great deal of sense and bode well for the future of the Valley's sustainability. Business leaders have learned valuable lessons from the past and are applying them to the current economic situation.

Factor 2: VCs Are More Diligent

As we have noted, the lifeblood of the Valley economy is venture capital (VC). At some point, investors need a liquidity event—IPOs and mergers allow VC firms to cash in their investments, hopefully at a profit.

It is expected that the impact of the current economic crisis on venture capital investments into Valley firms will not be as serious as what occurred in the aftermath of the Dot-Com Bust, as the level of due diligence on the part of investors has been much higher this time around.[150]

Yet the financial and economic situation is significantly impacting the flow of capital into the Valley. Silicon Valley IPOs and mergers declined dramatically in 2008. In 2008, only seven companies in the Valley held IPOs, compared to 76 in 2007. This meant an infusion of only $551 million of cash back into VC firms, compared to $6.8 billion in 2007.

The financial crisis has also resulted in longer time spans for liquidity events that repay Valley VC firms in 2008. For the seven firms that held IPOs in 2008, the median age was a record-breaking 8.3 years. The implication of this is clear: a decline in the frequency of new start-ups in the Valley as well as lower returns for VC firm partners.[151]

As a result, VC investors have relied on mergers and acquisitions (M&A) as the primary avenue to liquidate their investments. While this is typical of the industry regardless of economic conditions, M&A activity, too, has declined dramatically by historical standards. Only 325 VC-backed mergers transpired in 2008, compared to 457 in 2007—the lowest figure since 1999.[152]

Frugality is clearly prized in the world of VCs in the Valley now.[153] VCs are seriously watching the burn rate of cash by the firms in their portfolios. In an effort to mitigate vulnerability to current and anticipated economic turmoil, two of the major Valley VC firms, Sequoia Capital (whose past successes include Google, Inc. and Yahoo! Inc.) and Benchmark Capital requested that the companies in their portfolios cut costs in their 2009 business plans. The idea is to give the firms a safety cushion to decrease the impacts of economic shocks.

Factor 3: Companies Are Conserving Cash

Some of the leaders of established companies have expressed guarded optimism about their industries and the Valley's economy in general.

Ivo Bolsens, the Chief Technology Officer of Xilinx, a specialty chip manufacturer in the Valley, said, "to the extent that all economic sectors will suffer from the [2008-2009] financial meltdown, it is likely that semiconductor companies will also be impacted adversely. In the short term, I do not expect serious consequences for the semiconductor and high tech businesses. Most of the leading high-tech companies (such as Xilinx) are in a good cash position. In the long run, capital intensive businesses such as chip manufacturing can be hurt."[154]

Factor 4: The Workforce Is Rebalancing

From a jobs perspective, the news was not all bad in 2008. Between July and August 2008, the Valley added a net of 300 jobs.[155] This is significant when we consider that the nation, the state, and other parts of the San Francisco Bay Area continued to lose jobs steadily since the beginning of that year. While most of the job cuts in the Valley occurred in the second half of 2008, some cleantech and life sciences firms have been actively hiring.

It is important to note that while layoffs continue in many of the Valley's major companies, Valley companies also report job openings, as they continue to invest in the development of new technology and new products.

What these firms are actually doing is rebalancing their workforce in order to meet the changing needs of the marketplace and emerging technology opportunities. This workforce dynamism is a key necessity of an ever-changing, innovation-driven economic region. This process appears to be alive and well so far in this negative business cycle, despite its intensity. Hopefully this trend will continue into 2009.

Factor 5: Cleantech Is Flourishing

The good news regarding the state of venture capital in the Valley in 2008 was in the cleantech sector.[156] Although investors are skittish about investing in general right now, the risk capital picture for the clean technology industry in the Valley looks promising.

In the third quarter of 2008, according to the Cleantech Group, the cleantech sector garnered $2.6 billion in VC financing worldwide.[157] Silicon Valley cleantech firms did well by obtaining $419 million of that figure. For all of 2008, cleantech businesses in the Valley received over $1 billion.

The top five cleantech investors included Google.org; Kleiner, Perkins, Caufield and Byers; and Khosla Ventures. The biggest Valley VC deal concluded in Q3-2008 involved the $200 million invested in the solar photovoltaic company Solopower.

The Valley's renewable power industry leaders were much relieved when the U.S. federal renewable energy tax credit was extended on October 1, 2008 as an add-on to the $700 billion bailout plan. The investment tax credit was extended for an additional eight years for solar power plants and for one year for wind-power projects. The tax credit for solar power is 30% of the cost of plants. Wind farm owners get a two-cent credit for every kilowatt hour of electricity they generate.[158]

This extension was critical to the success of the cleantech energy industry, as the tax credits were set to expire at the end of 2008. Without tax credits, many of the nation's large-scale alternative energy projects would not be economically viable. Failure of the tax credit extension would have been a major blow to the industry, which has the promise of reshaping and revitalizing not only the Valley's economy but the nation's economy as well.

Survival and Beyond

These five factors—smarter start-ups, more diligent VC investing, more aggressive cash management, workforce rebalancing, and the flourishing cleantech sector—all suggest that the Valley is actively adapting to the tough economic realities of 2009. Unlike other regions that bemoan the collapse of their economies, industries that resist redirection in the face of rapidly changing market conditions, or workers who are complacent in an increasingly competitive global economy, the Valley's survival and prosperity over time is rooted in its culture that embraces change and aggressively adapts to the new realities. Throughout the Valley, we see changing business practices, intense market focus, and greater due diligence in investments—all with the goal of maintaining the Valley's viability in the rapidly changing global economy.

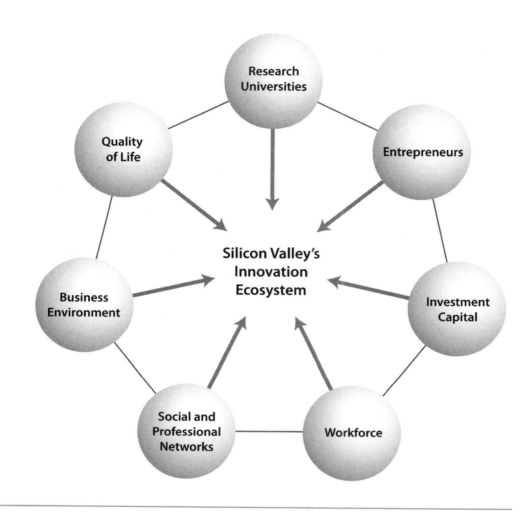

The Seven Key Elements of Silicon Valley's Innovation Ecosystem

Can Silicon Valley Be Cloned?

Clearly Silicon Valley's economy is unique, and given all the elements and relationships and synergies that make Silicon Valley tick, it would be extremely difficult if not impossible to clone or replicate the Valley in any other place. Silicon Valley is far more than just a collection of high-tech companies and high net-worth investors. The Valley is blessed with physical beauty; a pleasant climate; great universities; a supportive business environment; and a high-energy, risk-taking, entrepreneurial culture that brings the best and brightest in the world together to create and innovate. Even bringing together similar ingredients and combining them in much the same way would not assure that the end result would have the same "flavor" as Silicon Valley.

Silicon Valley is the preeminent innovation economy in the world. Time and again, it has overcome major economic ups and downs, technological upheavals, and brutal competition. The Valley's resilience is the foundation of its long-term sustainability. Looking at the region's economy as an ecosystem, we have seen that the Valley's success is based on seven key elements that adapt and evolve in the face of constant change: research universities, entrepreneurs, investment capital, talented workers, social and professional networks, business environment and quality of life.

The Most Important Factor: Multiple Networks

From our perspective, the most important and unique element of the Valley's ecosystem is its social and professional networks. The interconnectedness, diversity and interdependence of people, businesses, institutions and stakeholders contribute to maintaining and enhancing the health of the Valley's economy. Stakeholders include world-class research universities, entrepreneurs, venture capitalists and angel investors, intellectual property attorneys, consultants, skilled workers and talented engineers.

Business plan competitions have four goals: to motivate researchers in the academic and business communities to come forward with ideas, to build the commercial skills of those researchers by bringing them together with business talent, to attract venture capital, and to identify service providers (such as patent attorneys, headhunters, and accountants) who can support entrepreneurial activities. In addition, each competition includes networking events designed to bring participants together with successful entrepreneurs and to facilitate access to supporting infrastructure, such as inexpensive office space and local patent licensing offices.

Source: The McKinsey Quarterly, *1999, Number 3*

Networks play a vital role in the movement of labor and information, and in the generation of ideas and innovations. They contribute greatly to building connections between the various elements of the Valley's ecosystem by fostering links among:

+ Newly minted firms and large established firms that serve as economic anchors
+ Companies of all sizes and the research universities and think tanks in the region and beyond
+ Immigrant entrepreneurs and the labor market and social and professional organizations
+ Immigrant entrepreneurs in Silicon Valley and their counterparts overseas
+ VCs in other countries and start-ups in Silicon Valley

Another important attribute of the Valley's ecosystem is that because of its history of innovation, acceptance and encouragement of diversity, and its culture of entrepreneurship, high-tech entrepreneurs find the Valley an excellent place to create new firms. Many of these high-tech start-ups succeed and grow and then spin off other firms, while some start-ups fail and close. The rapid pace of the birth and death of companies is part of the dynamic of an innovation ecosystem. It is just the same in a biological ecosystem: Only the strong survive. The long-term health of the region's economy ultimately depends on the ongoing regenerative process of new business creation. Another way of looking at the Valley's ecosystem is to see it as a prime business incubating region, another attribute that sets the Valley apart from other high-tech regions of the world.

Silicon Valley has over the last three decades become the model to follow for successful high-tech regions elsewhere in the world. A large number of "Silicon Somewheres" are really industrial parks for silicon-chip manufacturing or biotechnology. They usually lack the fundamental elements and attributes of the Silicon Valley innovation ecosystem.

Other successful innovation ecosystems in India (Bangalore), Israel, and Taiwan (Sinchu) have benefited from the involvement of expatriates returning home from Silicon Valley.

Recommendations for Other Regions

It may not be possible to clone Silicon Valley's innovation economy, but there are many useful lessons to be learned from the region's innovation ecosystem that can be applied to other places seeking to emulate the Valley's success. Here are our recommendations for other regions that aspire to develop successful and sustainable innovation economies based on the Silicon Valley model:

- Capitalize on their region's unique resources, talents and opportunities while developing virtual global relationships.
- Focus on improving and marketing the region's quality of life as well as its business environment.
- Foster high-quality educational systems and strong connections between universities and businesses. These are as important to a region's long-term success as seaports, airports, paved highways and digital highways.
- Develop a nurturing environment for entrepreneurs and encourage innovative risk-taking. It is important to recognize the successes of local entrepreneurs and not dwell on their failures. Help make them role models.
- Encourage young people to pursue their passions. Teach them the skills of entrepreneurship. Give them lots of opportunities to gain practical business experience. Business plan competitions and business forums are excellent tools for building "farm teams" (as the junior leagues in sports are called) to nurture next-generation entrepreneurs.

- Help develop extensive social and professional networks that create connections between research universities, venture capitalists, entrepreneurs, specialized lawyers, accountants, consultants, scientists and engineers.
- Help attract the best scientific, technical, and managerial talents from all over the world to the research universities and think tanks in the region.
- Develop technology-transfer offices at regional universities that encourage entrepreneurial faculty and graduate students to start new businesses in the region. Provide facilities like incubators and innovation parks around universities to support new and early-stage companies.
- Invite U.S. and European angel investors and venture capitalists to forums and conferences where regional start-ups can show off their investment potential for foreign investors. This is likely to result in an inflow of risk capital from abroad, as well as to gradually encourage the development of a culture of risk-taking in the region.
- Help develop angel capital networks in the region. This will encourage the investment of early stage capital from regional or local sources as well as enhance the culture of risk-taking. Establish links with U.S. angel groups such as the Keiretsu Forum and the Band of Angels. Use these groups as models.
- Develop hybrid VC and angel investment funds that combine government and private capital to support regional start-up companies with local and regional sources of funds.
- Ensure that intellectual properties (IPs) are protected against piracy.

As we have observed, Silicon Valley's success as an innovation economy is rooted in the way in which it has evolved as a healthy, resilient, adaptable and sustainable innovation ecosystem. Silicon Valley can be emulated by creating, fostering and nurturing a region's own set of key elements—when that is done, a healthy innovation ecosystem will very likely evolve.

Looking Forward

As we have discussed, Silicon Valley's innovation ecosystem is currently facing a major challenge from the global recession—the worst economic crisis since the Great Depression of the 1930s. Earlier we assessed the impact of this economic debacle on the Valley. Our conclusion is that the Valley is not immune to economic debacles such as the current one. Not surprisingly, businesses are laying off workers and looking for cost savings. However, these layoffs are not dramatic in a macro sense, as unemployment levels are reasonable compared to other subregions of the San Francisco Bay Area.

There are three important things to note about the adjustments being made by the key players in the Valley:

- VC money, although tight and carefully weighed, is pouring into what I would call the "next big thing" in the Valley: the cleantech sector.
- Today's start-ups, unlike those in the pre Dot-Com era, are careful with their money and focus on viability, rather than speed of gaining market share. Acceptance of high money "burn rates" is over and is being replaced by an emphasis on survival and frugality.
- The established flagship companies are terminating their employees but at the same time rebalancing their workforce in order to meet the changing needs of the marketplace and the need for sustaining innovation.

Understanding and emulating Silicon Valley, one of the premier innovation economies of the world, is of paramount national importance for all of us at this time. Silicon Valley provides us with an excellent model of an innovation ecosystem that can be emulated in the U.S. and abroad in order to enhance economic vitality of regions and nations. The lessons from the Valley can be adapted for differences in geography and culture.

Looking forward, it is quite important for all of us to accept the fundamental tenet that in the 21st century, innovation is not a luxury. It is a necessity for our economic survival and prosperity.

Introduction

1. A new word, *Siliconia*, represents the appropriation of names beginning with *Silicon* by areas outside Silicon Valley.

2. H. Bahrami and S. Evans, *Flexible Recycling and High Technology Knowledge Entrepreneurship in Understanding Silicon Valley—The Anatomy of an Entrepreneurial Region*, ed. Martin Kenney (Stanford University Press, 2000), 169.

Chapter 1: A History of Innovation

3. Timothy J. Sturgeon, "How Silicon Valley Came to Be," in *Understanding Silicon Valley: Anatomy of an Entrepreneurial Region*, ed., Martin Kenney (Stanford University Press, 2000).

4. National Museum of American History, http://smithsonianchips.si.edu/schreiner/hoefler.htm

5. Joint Venture: Silicon Valley Network, *2007 Silicon Valley Index*.

6. Reed Albergotti, "The Most Inventive Towns in America," *Wall Street Journal*, 22-23 July 2006 and *2008 Index of Silicon Valley*, Joint Venture: Silicon Valley Network.

7. The campus is to this day often referred to casually as "the Farm."

8. Timothy J. Sturgeon, op cit.

9. Ibid.

10. At its founding, the Federal Telegraph Company was named the Poulsen Wireless Telephone and Telegraph Company.

11. Sturgeon, op. cit.

12. Carolyn Tajna, *Fred Terman—The Father of Silicon Valley* (Stanford University, 1995).

13. Ibid.

14. Sturgeon, op. cit.

15. Stephen Adams, "Regionalism in Stanford's Contribution to the Rise of Silicon Valley," *Oxford Journals*, 2003.

16. Sturgeon, op. cit.

17. Tajna, op cit.

18. SiliconValleyOnline.org

19. With John Bardeen and Walter Brattain.

20. http://72.14.253.104/search?q=cache:fxNTQ3talBgJ:www.london.edu/ assets/documents/PDF/LourdesSosaPaper.pdf+%22shockley+transisto rs%22&hl=en&ct=clnk&cd=16&gl=us

21. Emilio J. Castilla, et al, "Social Networks in Silicon Valley," Fairchild Semiconductor, IEEE Virtual Museum, and Silicon Valley Networks Analysis Project, Stanford University, 2000.

22. http://www.sjsu.edu/depts/PoliSci/faculty/christensen/sj_history.htm

23. http://www.gcase.org/article_10016.html

24. SiliconValleyOnline.org

25. K. Southwick, *High Noon: The Inside Story of Scott McNealy and the Rise of Sun Microsystems* (John Wiley & Sons, 1999).

26. Edward B. Fiske, "Stanford Takes 'Audacious' Steps to Diversify at 102," *New York Times*, February 11, 1987.

Chapter 2: The Silicon Valley Economy

27. Tapan Munroe, *Dot-Com to Dot-Bomb: Understanding the Boom, Bust, and the Resurgence* (Moraga Press, 2004).

28. "World Knowledge Competitiveness Index, 2005," Center for International Competitiveness, http://www.cforic.org

29. Michael Porter, *The Competitive Advantage of Nations* (Basic Book, 1990).

30. Alfred Marshall, *Principles of Economics*, 1890.

31. Cortwight and Meier, High Tech Specialization: A Comparison of High Technology Centers (Brookings Institute, 1998).

32. http://www.brookings.edu/es/urban/cortright/specialization.pdf

33. U.S. Bureau of Economic Analysis, Moody's Economy.com

34. Steven Kosiak and Elizabeth Heeter, "Post-Cold War Defense Spending Cuts: A Bipartisan Decision," Center for Strategic and Budgetary Assessments, August 31, 2000.

35. Ibid.

36. U.S. Bureau of Economic Analysis, Moody's Economy.com

37. "Personal Computer Sales Still High," *Purchasing*, April, 1995.

38. CERN originally stood for the Conseil Européen pour la Recherche Nucléaire. CERN's name was changed in 1954 to the European Council for Nuclear Research but the acronym remains in common use. The World Wide Web began as a project at CERN in 1990. The first Web server at CERN was a NeXT cube made by Steve Jobs' NeXT Computer, Inc., a company based in Silicon Valley.

39. U.S. Bureau of Labor Statistics, Moody's Economy.com

40. Moody's Economy.com

41. Ibid.

42. In her book, *Closing the Innovation Gap: Reigniting the Spark of Creativity in a Global Economy* (McGraw-Hill, 2009, p. 36), Judith Estrin suggests that the various inhabitants of an economic ecosystem such as scientists, administrators, engineers, and educators can be categorized into three principal communities: research, development, and application. This is a variation of the framework we are presenting in this book.

Chapter 3: Key Elements

43. In this connection it is important to review the following books: James. F. Moore, *The Death of Competition: Leadership and Strategy in the Age of Business Ecosystems* (Harper Business, 1996); Marco Iansiti and Roy Levien, *The Keystone Advantage: What the Dynamics of Business Ecosystems Mean for Strategy, Innovation, and Sustainability* (Harvard Business School Press, 2004).

44. University of California campus locations include Berkeley, San Francisco, Merced, Santa Cruz, Los Angeles, San Diego, Irvine, Riverside, Santa Barbara, and Davis.

Chapter 4: Research Universities

45. http://www.universityofcalifornia.edu/ecnomy/clusters.html

46. Christophe Lecuyer, Bettina Horne-Muller, and Cherisa Yarkin, "Contributions by University of California Scientists to the State's Electronics Manufacturing Industry," Working Paper 07-03, Inter-University Cooperative Research Program, University of California, August 1, 2007, 33.

47. Ibid.

48. Ibid.

49. T. Bradshaw, Tapan Munroe, and Mark Westwind, "Economic Development Via University-Based Technology Transfer: Strategies for Non-Elite Universities," *Journal of Technology Transfer*, October 2003.

50. Tapan Munroe, "New Heyday Under Way in Silicon Valley," *Contra Costa Times*, April 6, 2006.

51. *2007 Silicon Valley Index*, op cit.

Chapter 5: Entrepreneurs

52. "Entrepreneurship is the process of identifying, developing, and bringing a vision to life. The vision may be an innovative idea, an opportunity, more simply a better way of doing something. The end result is the creation of a new venture, formed under conditions of risk and considerable uncertainty." The Entrepreneurship Center at Miami University of Ohio.

53. Ibid, pages 98-101.

54. Chong-Moon Lee, et al, *The Silicon Valley Edge: A Habitat for Innovation and Entrepreneurship* (Stanford University Press, 2000).

55. Digital-Lifestyles.com, December 10, 2007.

56. Tapan Munroe, "Two Examples of Innovation at Work," *Oakland Tribune*, February 12, 2009.

57. BetterPlace.com

58. Wadhwa, V., et. al., *America's New Immigrant Entrepreneurs* (Duke University School of Engineering, January 2007).

59. Ibid.

Chapter 6: Investment Capital

60. http://wsbe2.unh.edu/files/Full%20Year%202006%2Media%20Release%20-%20March%202007.pdf

61. "Venture Support Systems Project: Angel Investors," *MIT Entrepreneurship Center*, February, 2000, 17.

62. John May and Elizabeth O'Halloran, *American Angel Investing*, (Charlottesville: University of Virginia, 2003).

63. http://www.equilar.com/NewsArticles/050106_WallSt-Journal.pdf

64. Center for Venture Research, University of New Hampshire.

65. http://www.clevelandfed.org/research/regional/Features/2006/february/angel_investing.cfm

66. http://entrepreneurship.mit.edu/Downloads/AngelReport.pdf

67. John May and Elizabeth O'Halloran, op. cit.

68. Ibid.

69. http://www.bandangels.com

70. http://www.bandangels.com

71. Don Dodge, "Don Dodge on Tthe Next Big Thing," (May 2007), http://dondodge.typepad.com/the_next_big_thing/2007/05/angel_capital_a.html

72. Ibid.

73. http://www.bandangels.com/entrepreneurs/index.php

74. Center for Venture Research, University of New Hampshire, Q1-2007.

75. http://www.equilar.com/NewsArticles/050106_WallSt-Journal.pdf

76. http://www.computerhistory.org/core/articles/remarkable_people.html

77. Leslie Berlin, *The Man Behind the Microchip: Robert Noyce and the Invention of Silicon Valley* (New York: Oxford University Press, 2005).

78. PriceWaterhouseCoopers and NVCA, *Money Tree Survey*, Q2- 2007.

79. http://nvca.org/def.html

80. Ibid.

81. http://www.vcinstitute.org/materials/galante.html

82. http://en.wikipedia.org/wiki/Venture_capital

83. http://en.wikipedia.org/wiki/Venrock

84. Emilio J. Castilla, et al., op cit.

85. Ibid.

86. http://www.nvca.org/pdf/2006seatattablepressreleasefinal-doc.pdf

87. Ibid.

88. *2007 Silicon Valley Index*, op. cit.

Chapter 7: Workforce

89. SiliconValleyOnline.org

90. *2007 Silicon Valley Index*, op. cit.

91. Ibid.

92. Annalee Saxenian, Chong-Moon Lee, et al, op. cit.

93. Jennifer Roberts, *San Jose Magazine*.

94. Tapan Munroe, op cit.

95. *San Jose Magazine*, November, 2007, 102, 104.

96. Tapan Munroe, "Success Requires Top-Flight Talent," *Contra Costa Times*, March 11, 2007.

97. Eric Schmidt and Hal Varian, "Google: Ten Golden Rules," MSNBC.com, *Newsweek*.

Chapter 8: Social and Professional Networks

98. Annalee Saxenian, op. cit.

99. Ibid.

100. Emilio J. Castilla, et al., op. cit.

Chapter 9: Business Environment

101. "Assessing the Competitiveness of California's Business Climate—Silicon Valley CEOs Speak Out," Silicon Valley Leadership Group, 2005.

102. Numbers in parenthesis represent the percentage of executives responding affirmatively to a particular issue.

103. www.svlg.net/CEOSurvey, Spring, 2008.

104. DataQuick, December, 2008.

105. Sue McCallister, *San Jose Mercury News*, December 18, 2008.

Chapter 10: Quality of Life

106. *Fortune Magazine*

Chapter 11: Resilience and Reinvention

107. Rachel Konrad, "Silicon Valley's Slump Saps Traditional Confidence," *Associated Press*, August 9, 2004.

108. Munroe Consulting, Inc.

109. Virginia I. Postrel "Resilience vs. Adaptation," *Forbes ASAP*, August 25, 1997.

110. Ibid.

111. http://www.dynamist.com/articles-speeches/asap/resilience.html

112. SiliconValleyOnline.org

113. Ibid.

114. Tapan Munroe, *Dot-Com to Dot-Bomb*, op. cit.

115. Ibid.

116. Ibid.

117. *2007 Silicon Valley Index*, op. cit.

118. "Global Web 2.0 Deals Up 14% In First Half of 2007 as Venture Capitalists Seek to Tap New Markets," Ernst & Young LLP, September 17, 2007.

119. Ibid.

120. http://www.economist.com/business/displaystorycfm?story_id=9230363

121. Cleantech Group, LLC.

122. *Cleantech* is an abbreviation of *clean technology*: knowledge-based products or services that improve operational performance, productivity, or efficiency while reducing costs, inputs, energy consumption, waste, or pollution. Its origin is the increased consumer, regulatory and industry interest in clean forms of energy generation—specifically, perhaps, the rise in awareness of global warming and the impact on the natural environment from the burning of fossil fuels.

123. Tapan Munroe, "Silicon Valley a Solar Leader," *Contra Costa Times*, November 5, 2006.

124. http://www.economist.com/business/displaystory.cfm?story_id= 9230363

125. PriceWaterhouseCoopers, *MoneyTree* Report

126. Ibid.

127. Cleantech Group, LLC.

128. Tapan Munroe, "Solar Power Ready to Shine," *Contra Costa Times*, November 18, 2007.

129. Matt Nauman, *Contra Costa Times*, October 8, 2008.

130. Andrea Quong, Red Herring.com, September 10, 2007.

131. www.Ausra.com

132. The World Economic Forum (WEF) is a Davos-based organization that hosts the well-known annual conference on the state global economy in Switzerland. WEF has been recognizing technology pioneers for the last several years.

133. Peter Carey, www. SiliconValley.com, November 11, 2007

Chapter 12: Challenges and Threats

134. Diane Rezendes Khirallah, Marianne Kolbasuk McGee, and Michelle Lodge, "Silicon Valley and the Culture of More," *Information Week*, September 25, 2000.

135. Diane Rezendes Khirallah, et. al., op. cit.

136. American Institute of Stress, 2005.

137. Tim Simmers, *Contra Costa Times*, October 22, 2007.

138. Diane Rezendes Khirallah, et. al., op. cit.

139. Timothy Roberts, *Silicon Valley/San Jose Business Journal*, May 30, 2007.

140. Carolyn Said, *San Francisco Chronicle*, May 30, 2007.

141. Ibid.

142. *Silicon Valley/San Jose Business Journal*, September 19, 2007.

143. Judith Estrin, *Closing the Innovation Gap: Reigniting the Spark of Creativity in a Global Economy* (McGraw-Hill, 2009).

144. Steve Hamm, "Is Silicon Valley Losing its Magic?," *Business Week*, January 12, 2009.

Chapter 13: Surviving Economic Storms

145. Scott Duke Harris, *San Jose Mercury News*, October 16, 2008.

146. Steve Johnson, MediaNews Group, December 18, 2008.

147. Associated Press, October 6, 2008.

148. Deborah Gage, "Intel Capital Finds Global Opportunities," *San Francisco Chronicle*, December 14, 2008.

149. Scott Duke Harris, op. cit.

150. Drew Voros, *Bay Area News Group*, December 18, 2008.

151. PriceWaterhouseCoopers, *MoneyTree* Report.

152. Ibid.

153. Tom Abate, "Startups Need to Stretch Their Dollars," *San Francisco Chronicle*, December 3, 2008.

154. Michael Liedtke, *Associated Press*, October 10, 2008.

155. California Employment Development Department, 2008.

156. The Cleantech industry includes an array of renewable technologies that include solar, wind, biomass, and energy efficiency technologies.

157. Cleantech Group, LLC.

158. Tapan Munroe, "Silicon Valley May Miss a Recession," *Bay Area News Group*, October 18, 2008.

Acknowledgements

I am thankful for the stimulating dialog I have had over the years on the role of innovation in regional economic development with my distinguished colleagues and friends around the world. They include Josep Pique (Chief Executive Officer of 22@Barcelona, a science park in Barcelona), Dr. Fransec Parallada (Professor at the Universitat Politecnica de Catalunya in Barcelona), several colleagues at La Salle University of Barcelona, and many science park executives throughout Spain. I have benefited also from many conversations with colleagues at the University of California, business associates, and journalists—too many to acknowledge individually here.

Special thanks are due to Dr. Felipe Romera, President of the Science Parks Association of Spain (APTE), for originally suggesting I write a book (which became *Silicon Valley: Ecology of Innovation*) two years ago. I am also grateful to APTE and Dr. Romera for permission to use significant portions of that book in the present one.

Let me also thank my coauthor Mark Westwind for his excellent contributions. In addition to writing Part 1 and the chapter on Investment Capital, he was the substantive editor of the manuscript and responsible for thematic continuity, as well as for preparation of the diagrams, tables and charts in the book. I am indebted to Kathe Grooms, publisher at Nova Vista Publishing, for her thoughtful guidance and support. Her meticulous editing has been an invaluable asset to us. Thanks also to Annette Krammer for her excellent design work, and to Molly Kelley for her editorial work. I would also like to thank my associate, Deborah Hall, for preparing the bibliography as well as supporting me in my research for this book.

Finally, I thank my wife, Astrid Munroe, for being a helpful companion and colleague on our many long trips to Spain to present numerous lectures over the past several years. As the principal reader of my writings over several decades, her contributions have always been valuable.

Tapan Munroe

About the Authors

Tapan Munroe, PhD

Tapan Munroe, PhD, is a recognized economist, author, speaker, and consultant in economic policy analysis. His expertise covers regional economics, innovation economics, and high-tech industry. His research and writings focus on the Silicon Valley and San Francisco Bay Area economies and the factors that support and sustain innovation and the formation of high-tech companies.

He is an affiliate of LECG, LLC, a worldwide consulting firm headquartered in Emeryville, California. He is a member of the Advisory Board of City National Bank in San Francisco; a member-Investor with Keiretsu Forum, the nation's largest business angel investment organization; a former trustee and member of the Investment Committee for the University of California at Merced Foundation; emeritus member of the Board of Directors of the Center for Pacific Rim Studies at the University of San Francisco; and a member of the University of California President's Board on Science and Innovation.

Tapan has served as the Chief Economist for the Pacific Gas & Electric Company, San Francisco, for more than a decade. Tapan is also a former president of the Bay Area Chapter of National Association of Business Economists, a former member of the National Petroleum Council Task Force on Oil Prices, former quarterly chairman for the Commonwealth Club of California and the former chairman of the Economics Committee for Edison Electric in Washington, D.C.

Tapan holds a PhD in Economics from the University of Colorado, where he was awarded a fellowship and membership in the Phi Kappa Phi Omicron Delta Epsilon. He is also a graduate of the University of Chicago Executive Training Program; a visiting scholar at the Massachusetts Institute of Technology,

Stanford University and the University of Augsburg in West Germany; an adjunct professor at the University of California, Berkeley and was a professor and Chairman at the Department of Economics at the University of the Pacific in Stockton for several years. He is currently a visiting faculty at the La Salle University in Barcelona. Tapan was the holder of the Kiriyama Distinguished Professorship for Asia Pacific Studies at the University of San Francisco.

A widely published author, Tapan is the author of the book *Dot-Com to Dot-Bomb—Understanding the Dot-Com Boom, Bust and Resurgence*. His second book, *Silicon Valley: The Ecology of Innovation* with coauthor Mark Westwind was published by the Science Parks Association of Spain (APTE) in November 2008. He has been a columnist on economic issues for the *San Francisco Examiner* and the *Journal of Corporate Renewal*. He is currently a columnist for several Media News papers in the San Francisco Bay Area, including the *Oakland Tribune* and the *Contra Costa Times*. He has been a commentator on both regional and national radio and TV news programs including KRON TV (channel 4), KGO TVT (Channel 7), CNBC in Los Angeles and New York, the Dow Jones Investors' Network and the Bloomberg News Service. Tapan can be contacted via tapan@tapanmunroe.com, and readers are invited to visit www.tapanmunroe.com.

Mark Westwind, MPA

Mark Westwind has more than 30 years experience working with businesses, government agencies and non-profit organizations. His company, Westwind Associates, is a professional and technical services firm in the San Francisco Bay Area. He has worked closely with Tapan Munroe on numerous publications and projects, including co-authoring *Silicon Valley: The Ecology of Innovation.*

Mark was the founding director of the Contra Costa Software Business Incubator and the founding Associate Director of John F. Kennedy University's Center for Entrepreneurship. During the Dot-Com era, Mark worked with several technology start-ups and produced a series of technology expositions in the East Bay. More recently, he served as the U.S. Associate of Canton Venture Capital Company, a Chinese VC firm, and as a mentor to U.S. and Chinese high-tech start-ups selected to participate in the annual World's Best Technology Showcase (his mentored company took the WBT 2005 Gold Award). He continues to serve as a technology consultant for the Contra Costa Small Business Development Center.

Mark is a co-founder of the Digital Safari Institute for Innovation in Education and a co-developer of a successful entrepreneurial training program and project-based framework for high school students. He is also the founder and producer of the Digital Safari Innovation Fair™—an annual exposition and business plan competition featuring "beyond the edge" innovations created by next-generation entrepreneurs.

Mark graduated from the University of California at Berkeley with a BA in Environmental Studies. He earned a Master of Public Administration degree from California State University–East Bay, and holds a Certificate of Entrepreneurship from John F. Kennedy University. He can be contacted via www.DigitalSafariInstitute.org.

Bibliography

A

Adams, Stephen. "Regionalism in Stanford's Contribution to the Rise of Silicon Valley." *Enterprise and Society*, 2003, 521-543.

Adams, Stephen. "Regionalism in Stanford's Contribution to the Rise of Silicon Valley." *Oxford Journals*, 2003.

Addison, Craig. *Silicon Shield: Taiwan's Protection against Chinese Attack.* Irvine, TX: Fusion Press, 2001.

Altman, Lawrence. "Semiconductor RAMs and Computer Mainframe Jobs." *Electronics*, August 28, 1972, 63-77.

Anandaram, Sanjay. "The Early Stage of U.S.-India Cross-Border Technology Start-up." *Asian Venture Capital Journal*, September 29, 2003.

Angel, David. "High Technology Agglomeration and the Labor Market: The Case of Silicon Valley." In *Understanding Silicon Valley: The Anatomy of an Entrepreneurial Region*, edited by Martin Kenney. Stanford, CA: Stanford University Press, 2000.

Angwin, Julia. "The Israeli Connection: Silicon Valley Promised Land for Startups." *San Francisco Chronicle*, September 10, 1997.

Arora, Ashish, and Alfonso Gambardella, eds. *From Underdogs to Tigers.* New York: Oxford University Press, 2005.

Arrow, Kenneth J. "Information and Economic Behavior." *Collected Papers.* Cambridge, MA: Harvard University Press, 1983, 136-52.

Asher, Norman, and Leland Strom. *The Role of the Department of Defense in the Development of Integrated Circuits.* Washington, DC: Institute for Defense Analysis, 1977.

Ashok, Sathya Mithra. "Seeking Silicon Valley." *CyberIndia Online*, March 5, 2004.

Atalla, M.M., E. Tannenbaum, and E.J. Scheibner. "Stabilization of Silicon Surfaces by Thermally Grown Oxides." *Bell System Technical Journal* 38 (1959): 749-83.

Autler, Gerald. "Global Networks in High Technology: The Silicon Valley-Israel Connection." Master's Thesis, Department of City and Regional Planning. University of California, Berkeley, 1999.

Aydlot, Phillippe, and David Keeble. *High Technology Industry and Innovative Environments: The European Experience*. New York: Routledge, Chapman, and Hall, 1988.

B

Bahl, Sheetal. "Is Offshore Demand Sustainable?" *Everest Research Institute*, July, 2005.

Bahrami, Homa, and Stuart Evans. "Flexible Recycling and High Technology Entrepreneurship." In *Understanding Silicon Valley: The Anatomy of an Entrepreneurial Region*, edited by Martin Kenney. Stanford, CA: Stanford University Press, 2000.

Bardham, Ashok Deo, and David K. Howe. "Transnational Social Networks and Globalization: The Geography of California's Exports." Working Paper No. 98-262, Fisher Center for Real Estate and Urban Economics, University of California, Berkeley, February 1998.

Barkley, David. "The Decentralization of High Technology Manufacturing to Non-Metropolitan Areas." *Growth and Change* 19 (1988): 12-30.

Bassett, Ross. *To the Digital Age: Research Labs, Start-Up Companies, and the Rise of MOS Technology*. Baltimore, MD: Johns Hopkins University Press, 2002.

Benjamin, G., and E. Sandles. "Angel Investors: Culling the Waters for Private Equity." *Journal of Private Equity* 1, no. 3 (Spring 1998): 41-59.

Benner, Chris. "Win the Lottery or Organize: Traditional and Non-Traditional Labor Organizing in Silicon Valley." *Berkeley Planning Journal*, 1998.

Berlin, Leslie. "Robert Noyce and Fairchild Semiconductor, 1957-1968." *Business History Review*, no. 75 (2001): 63-102.

Berlin, Leslie. "Entrepreneurship and the Rise of Silicon Valley: The Career of Robert Noyce, 1956-1990." Ph.D dissertation, Stanford University, 2001.

Berlin, Leslie. *The Man behind the Microchip: Robert Noyce and the Invention of Silicon Valley.* New York: Oxford University Press, 2005.

Biswas, Smita. "Leading the VC Wave in India." *Siliconindia.* May-June, 1998.

Black, Bernard S., and Ronald J. Gilson. "Venture Capital and the Structure of Capital Markets: Banks Versus Stock Markets." *Journal of Financial Economics,* no. 47 (1997): 243-77.

Booz Allen Hamilton. "Foreign Investment in Bay Area Bioscience." A survey prepared for the Bay Area Bioscience Center, 1991.

Bradshaw, Ted, Tapan Munroe and Mark Westwind. "Economic Development via University-based Technology Transfer: Strategies for Non-Elite Universities." *Journal of Technology Transfer,* October, 2003.

Bresnaham, Timothy, and Manuel Trajtenberg. "General Purpose Technologies, 'Engines of Growth.'" *Journal of Econometrics,* no. 65 (1995): 83-108.

Breznitz, Danny. "High Tech, Innovation, and the Periphery: The Role of the Military in the Israeli Success." *PreCIS* 11, no. 4 (2000): 8-10.

Buderi, Robert. *The Invention That Changed the World.* New York: Simon and Schuster, 1996.

Burton, Lawrence, and Jack Want. "Issue Brief: How Much Does the U.S. Rely on Immigrant Engineers?" Division of Science Resource Studies, National Science Foundation, Arlington, Virginia, February 1, 1999.

Bygrave, W., and J. Timmons. *Venture Capital at the Crossroads.* Cambridge, MA: Harvard Business School Press, 1992.

Bylinsky, Gene. "California's Great Breeding Ground for Industry." *Fortune,* June 1974, 129-35, 216-24.

Bylinsky, Gene. "How Intel Won Its Bet on Memory Chips." *Fortune,* November 1973.

C

Cairncross, Frances. *The Death of Distance: How the Communication Revolution Will Change Our Lives.* Boston, MA: Harvard Business School Press, 1997.

Calkins, Robert, and Walter Hoadley. *An Economic and Industrial Survey of the San Francisco Bay Area.* California Estate Planning Board, 1941.

Callon, Scott. "Different Paths: The Rise of Taiwan and Singapore in the Global Personal Computer Industry." Asia/Pacific Research Center, Stanford University, January, 1995.

Castells, Manuel. *The Rise of the Network Society.* London: Blackwell Publishers, 1996.

Castells, Manuel, and Peter Hall. *Technopoles of the World: The Making of Twenty-First Century Industrial Complexes.* London: Routledge, 1994.

Castilla, Emilio J., et al. "Social Networks in Silicon Valley." *Silicon Valley Network Analysis Project.* Stanford, CA: Stanford University Press, 2000.

Chan, Wai-Chan, Martin Hirt, and Stephen Shaw. "The Greater China High-Tech Highway." *McKinsey Quarterly 4,* 2002.

Chang, Shirley L. "Causes of Brain Drain and Solutions: The Taiwan Experience." *Studies in Comparative International Development,* 27, no. 1 (1992): 27-43.

Chen, Edward K. Y. "The Electronics Industry of Hong Kong: An Analysis of Its Growth." Master's Thesis, University of Hong Kong, 1971.

Cheng, Lu-Lin, and Gary Gereffi. "The Informal Economy in East Asian Development." *International Journal of Urban and Regional Research* 18, no. 2 (1994): 194-219.

Chiruvolu, Ravi. "TechTalk: About That India Recommendation…" *Venture Capital Journal,* November 1, 2003.

Christensen, Clayton M. *The Innovator's Dilemma: When New Technologies Cause Great Firms to Fail.* Boston, MA: Harvard Business School Press, 2000.

Chugani, Michael. "Chinese IT Professionals Flock Back Home." *South China Morning Post,* January 12, 2002.

Cohen, Steven, and John Zysman. *Manufacturing Matters: The Myth of the Post-industrial Economy.* New York: Basic Books, 1987.

Cohen, Y., and Y. Haberfeld. "Self-Selection and Return Migration: Israeli-Born Jews Returning Home from the United States During the 1980s." *Population Studies* 55, no. 1 (2001): 79-91.

Collaborative Economics. *Innovative Regions: The Importance of Place and Networks in the Innovative Economy.* Pittsburgh, PA: The Heinz Endowments, 1999.

Cortright, Joseph, and Heike Meier. "High Tech Specialization: A Comparison of High Technology Centers." Survey Series. Brookings Institution, January 2001.

Cox, Gail D. "A Valley of Conflicts." *National Law Journal,* June 20, 1988, 1-46, 48-49.

D

Dataquest. "Dataquest 20 Years: The Great Indian Software Revolution." 2002.

DeFontenay, Catherine, and Erran Carmel. "Israel's Silicon Wadi: The Forces behind Cluster Formation." In *Building High-Tech Clusters: Silicon Valley and Beyond,* edited by T. Brenahan and A. Cambardella, eds. Cambridge: Cambridge University Press, 2004.

Delbecq, Andre, and Joseph Weiss. "The Business Culture of Silicon Valley: Is It a Model for the Future?" In *Regional Cultures, Managerial Behavior and Entrepreneurship,* edited by Joseph Weiss. New York: Quorum Books, 1988.

DeVol, Ross C. *America's High-Tech Economy.* Santa Monica, CA: Milken Institute, 1999.

Dobbs, Gordon B., and Craig E. Wollner. *The Silicon Forest: High Tech in the Portland Area, 1945-1986.* Portland, OR: Oregon Historical Society Press, 1990.

Dorfman, Nancy S. "Route 128: The Development of a Regional High Technology Economy," *Research Policy* 12 (1983): 299-316.

E

Engstrom, Therese. "Little Silicon Valley." *Technology Review*, January 1987: 24-32.

Estrin, Judy. *Closing the Innovation Gap: Reigniting the Spark of Creativity in a Global Economy*. New York: McGraw-Hill, 2008.

F

Fallick, Bruce, Charles A. Fleischman, and James B. Rebitzer. "Job Hopping in Silicon Valley: The Micro-Foundation of a High Technology Cluster." Working Paper, Weatherhead School of Management, Case Western Reserve University, 2003.

Ferguson, Charles H. "Computers and the Coming of the U.S. Keiretsu." *Harvard Business Review*, no. 4 (1990): 55-70.

Fiske, Edward B. "Stanford Takes 'Audacious' Steps to Diversify at 102." *New York Times*. February 11, 1987.

Florida, Richard L., and Martin Kenney. "Venture Capital and High-Technology Entrepreneurship." *Journal of Business Ventures*, no. 3 (1987): 301-19.

Florida, Richard, and Martin Kenney. "Organizational Transplants: The Transfer of Japanese Industrial Organization to the U.S." *American Sociological Review* 56, no. 3 (June 1991): 381-98.

Freeman, John. "Venture Capital as an Economy of Time." In *Corporate Social Capital*, edited by R. Leenders and S. Gabby. New York: Addison Wesley, 1999, 400-19.

Freeman, Linton C. "Centrality in Social Networks: Conceptual Classification." *Social Networks 1* (1979): 215-39.

Freiberger, Paul, and Michael Swaine. *Fire in the Valley: The Making of the Personal Computer*. New York: McGraw-Hill, 2000.

G

Gage, Deborah. "Intel Capital Finds Global Opportunities." *San Francisco Chronicle*, December 14, 2008.

Garnsey, Elizabeth, and Helen Lawton Smith. "Proximity and Complexity in the Emergence of High Technology Industry: The Oxbridge Comparison," *Geoforum* 29, no. 4 (1988): 433-50.

Gillmor, Dan. "High-Tech Blooming in Israel." *San Francisco Chronicle*, April 1, 1998.

Gold, Thomas B. *State and Society in the Taiwan Miracle*. Armonk, NY: Sharpe, 1986.

Golding, Anthony. "The Semiconductor Industry in Britain and the United States: A Case Study in Innovation, Growth, and the Diffusion of Technology." Doctoral thesis, University of Sussex, 1971.

Gonzales-Benito, Javier, Stuart Reid, and Elizabeth Garnsey. "The Cambridge Phenomenon Comes of Age." *Research Papers in Management Studies*, WP 22/97. Cambridge: Judge Institute of Management Studies, 1977.

Graham, Margaret, and Alec Shuldiner. *Corning and the Craft of Innovation*. New York: Oxford University Press, 2001.

Granovetter, Mark. "The Economic Sociology of Firms and Entrepreneurs." In *The Economic Sociology of Immigration: Essays on Networks, Ethnicity and Entrepreneurship*, edited by Alejandro Portes. New York: Russell Sage, 1995.

Grove, Andrews S. *Only the Paranoid Survive*. New York: Currency/Doubleday, 1996.

H

Hauben, Michael, and Ronda Hauben. *Netizens: On the History and Impact of Usenet and the Internet*. Los Alamitos, CA: IEEE Computer Society Press, 1997.

Hayes, Dennis. *Behind the Silicon Curtain: The Seduction of Work in a Lonely Era*. Cambridge, MA: South End Press, 1989.

Heinrich, Thomas. "Cold War Armory: Military Contracting in Silicon Valley." *Enterprise and Society* 3 (2002): 247-284.

Hoefler, Don C. "Silicon Valley, USA." *Electronic News*, January 1, 1971.

I

Iansiti, Marco, and Roy Levien. *The Keystone Advantage: What the Dynamics of Business Ecosystems Mean for Strategy, Innovation, and Sustainability.* Cambridge, MA: Harvard Business School Press, 2004.

J

Jackson, Tim. *Inside Intel: Andy Grove and the Rise of the World's Most Successful Chip Company.* New York: Dutton, 1997.

Johnston, Moira. "High Tech, High Risk, and High Life in Silicon Valley." *National Geographic* 162, no. 4 (October 1982): 459-76.

Joint Venture: Silicon Valley Network. *The 2004 Index of Silicon Valley.* San Jose, California, 2004.

K

Kaplan, David. *The Silicon Boys and Their Valley of Dreams.* New York: William Morrow and Co., 1999.

Kenney, Martin, and Richard Florida. "Venture Capital in Silicon Valley: Fueling New Firm Formation." In *Understanding Silicon Valley: Anatomy of an Entrepreneurial Region*, edited by Martin Kenney. Stanford, CA: Stanford University Press, 2000.

Kenney, Martin, and Urs von Burg. "Technology and Path Dependence: The Divergence Between Silicon Valley and Route 128." *Industrial and Corporate Change* 8, no. 1 (1999): 67-103.

Kenney, Martin, and Urs von Burg. "Institutions and Economies: Creating Silicon Valley." In *Understanding Silicon Valley: Anatomy of an Entrepreneurial Region*, edited by Martin Kenney. Stanford, CA: Stanford University Press, 2000.

Kenney, Martin, ed. *Understanding Silicon Valley: The Anatomy of an Entrepreneurial Region*. Stanford, CA: Stanford University Press, 2000.

Khavul, Susanna. "The Emergence and Evolution of Israel's Software Industry." Center for New and Emerging Markets, London Business School, June 2003.

Khirallah, Diane Rezendes, Marianne Kolbasuk and Michelle Lodge, "Silicon Valley and the Culture of More." Clarity Seminars, www.clarityseminars.com/info_week.html (accessed September 25, 2000).

Kleiman, Herbert. "The Integrated Circuit: A Case Study of Production Innovations in the Electronics Industry." Ph.D. dissertation, George Washington University, 1966.

Kleiner, Eugene, et al. *Fairchild Semiconductor*. IEEE Virtual Museum.

Konrad, Rachel. "Silicon Valley's Slump Saps Traditional Confidence." *Associated Press*, August 9, 2004.

Kosiak, Steven, and Elizabeth Heeter. "Post-Cold War Defense Spending Cuts: A Bipartisan Decision." *Center for Strategic and Budgetary Assessments*. August 31, 2000.

Krantzler, Mel, and Patricia Biondi Krantzler. *The High Cost of the High-Tech Dream: Down and Out in Silicon Valley*. New York: Prometheus Books, 2002.

Krugman, Paul. *Development, Geography and Economic Theory*. Ohlin Lectures, vol. 6. Cambridge, MA: MIT Press, 1995.

Krugman, Paul. "What's New about the New Economic Geography?" *Oxford Review of Economic Policy* 14, no. 2 (1996).

Kvamme, Floyd. "Life in Silicon Valley: A First-Hand View of the Region's Growth." In *The Silicon Valley Edge*, edited by C.-M. Lee, et. al. Stanford, CA: Stanford University Press, 2000.

L

Lécuyer, Christophe. "Making Silicon Valley: Engineering Culture, Innovation, and Industrial Growth, 1930-1970." Ph.D. dissertation, Stanford University, 1999.

Lécuyer, Christophe. "Fairchild Semiconductor and Its Influence." In *The Silicon Valley Edge*, edited by C.-M. Lee, et. al. Stanford, CA: Stanford University Press, 2000.

Lécuyer, Christophe. "What Do Universities Really Owe Industry? The Case of Solid State Electronics at Stanford." *Minerva*, no. 43 (2005): 51-71.

Lécuyer, Christophe. *Making Silicon Valley: Innovation and the Growth of High Tech, 1930-1970*. Cambridge, MA: MIT Press, 2006.

Lécuyer, Christophe, Bettina Horne-Muller, and Cherisa Yarkin. "Contributions by University of California Scientists to the State's Electronics Manufacturing Industry." Inter-University Cooperative Research Program. University of California, Berkeley, August 1, 2007.

Lee, Chong-Moon, et. al., eds. *The Silicon Valley Edge: A Habitat for Innovation and Entrepreneurship*. Stanford, CA: Stanford University Press, 2000.

Leonard-Barton, Dorothy. *Wellsprings of Knowledge: Building and Sustaining the Sources of Innovation*. Cambridge, MA: Harvard Business School Press, 1995.

Leone, Anthony, Jose Vamos, Robert Keeley, and William F. Miller. *A Survey of Technology-Based Companies Founded by Members of the Stanford University Community*. Stanford, CA: Office of Technology Licensing, Stanford University, 1993.

Leslie, Stuart. "How the West Was Won: The Military and the Making of Silicon Valley." In *Technological Competitiveness*, edited by W. Aspray. Piscataway, NJ: IEEE Press, 1993.

Leslie, Stuart. "Regional Disadvantages: Replicating Silicon Valley in the New York's Capital Region." *Technology and Culture* 42 (2001): 236-264.

Leslie, Stuart. "Blue Collar Science: Bringing the Transistor to Life in the Lehigh Valley." *Historical Studies in the Physical and Biological Sciences* 42 (2001): 71-113.

Leslie, Stuart, and Robert Kargon. "Electronics and the Geography of Innovation in Post-War America." *History and Technology* 11 (1994): 217-231.

Leslie, Stuart, and Robert Kargon. "Selling Silicon Valley: Frederick Terman's Model for Regional Advantage." *Business History Review* 70 (1996): 435-472.

Levin, Richard. "The Semiconductor Industry." In *Government and Technical Progress*, edited by R. Nelson. Oxford: Pergamon Press, 1982.

Light, Ivan, and Edna Bonacich. *Immigrant Entrepreneurs: Koreans in Los Angeles, 1965-1982*. Berkeley: University of California Press, 1998.

Linde, Lucinda, and Alok Prasad. "Venture Support Systems Project: Angel Investors." MIT Entrepreneurship Center, Cambridge, Massachusetts, February, 2000.

Linvill, John, and Lester Hogan. "Intellectual and Economic Fuel for the Electronic Revolution." *Science*, March 1, 1977, 1107-1113.

Lösch, August. *The Economics of Location*. Translated from the Second Revised Edition (1959) by William H. Woglom. New York: John Wiley & Sons, 1967.

Lowen, Rebecca. *Creating the Cold War University: The Transformation of Stanford*. Berkeley, CA: University of California Press, 1997.

Lowood, Henry. *From Steeple of Excellence to Silicon Valley*. Varian Associates, Palo Alto, California, 1989.

M

Mackun, Paul. "Silicon Valley and Route 128: Two Faces of the American Technopolis." University of California, Berkeley, http://netvalley.com/archives/mirrors/sv&128.html.

Malone, Michael. *The Big Score: The Billion Dollar Story of Silicon Valley*. New York: Doubleday Books, 1985.

Malone, Michael. *The Microprocessor: A Biography*. Santa Clara, CA: Springer-Verlag, 1995.

Marshall, Alfred. *Principles of Economics*. Reprint, 1890, Amherst, NY: Prometheus Books, 1997.

Mathews, John A. "A Silicon Valley of the East: Creating Taiwan's Semiconductor Industry." *California Management Review* 39, no. 4 (Summer, 1997): 26-54.

Mathews, John. "A Silicon Island of the East: Creating a Semiconductor Industry in Singapore." *California Management Review* 41, no. 2 (1999): 55-78.

Mathews, John, and Dong Sung Cho. *Tiger Technologies: The Creation of a Semiconductor Industry in East Asia.* Cambridge University Press, 2000.

May, John, and Elizabeth O'Halloran. *American Angel Investing.* Charlottesville, VA: University of Virginia. 2003.

McKendrick, David, Richard F. Doner, and Stephen Haggard. *From Silicon Valley to Singapore: Location and Competititve Advantage in the Hard Disk Drive Industry.* Stanford, CA: Stanford University Press, 2000.

Miller, Roger, and Marcel Côté. "Growing the Next Silicon Valley." *Harvard Business Review*, July-August 1985, 14-23.

Miller, Roger-Emile. *Growing the Next Silicon Valley: A Guide for Successful Regional Planning.* Lexington,, MA: Lexington Books, 1987.

Moore, James F. "The Death of Competition: Leadership and Strategy in the Age of Business Ecosystems." *Harper Business*, 1996.

Mowery, David C., and Nathan Rosenberg. "The U.S. National Innovation System." In *National Systems of Innovation*, edited by Richard R. Nelson. New York: Oxford University Press, 1993.

Mufson, Steven. "The Information Highway—From Texas to Beijing." *Washington Post*, June 16, 1998, A1.

Munroe, Tapan. "Silicon Valley a Solar Leader." *Contra Costa Times*, November 5, 2006.

Munroe, Tapan. "Success Requires Top Flight Talent." *Contra Costa Times*, March 11, 2007.

Munroe, Tapan. 2003. "New Heyday Under Way in Silicon Valley." *Contra Costa Times*, April 6, 2006.

Munroe, Tapan. *Dot-Com to Dot-Bomb: Understanding the Boom. Bust, and the Resurgence.* Moraga, CA: Moraga Press, 2004.

N

Nohria, Nitin. "Information and Search in the Creation of New Business Ventures: The Case of the 128 Venture Group." In *Networks and Organizations: Structure, Form, and Actions,* edited by N. Nohria and R. Eccles. 240-61. Boston: Harvard Business School Press, 1992.

Norberg, Arthur L. "The Origins of the Electronics Industry on the Pacific Coast." *Proceedings of the IEEE* 64, no.9 (September 1976): 1314-22.

O

O'Mara, Margaret Pugh. *Cities of Knowledge: Cold War Science and the Search for the Next Silicon Valley.* Princeton, NJ: Princeton University Press, 2004.

P

Piore, Michael, and Charles Sabel. *The Second Industrial Divide: Possibilities for Prosperity.* New York: Basic Books, 1984.

Porter, Michael. *The Competitive Advantage of Nations.* New York: Free Press, 1990.

Porter, Michael.. "Clusters and the New Economics of Competition." *Harvard Business Review,* November-December 1998, 77-90.

Portes, Alejandro, ed. *The Economic Sociology of Immigration: Essays on Networks, Ethnicity, and Entrepreneurship.* New York: Russell Sage, 1995.

Portes, Alejandro, ed. "Global Villagers: The Rise of Transnational Communities." *The American Prospect,* March-April, 1996, 74-77.

Postrel, Virginia I. "Resilience vs. Adaptation," *Forbes ASAP,* August 25, 1997.

Powell, Walter, and Laurel Smith-Doerr. "Networks and Economic Life." In *The Handbook of Economic Sociology,* edited by Neil J. Smelser and Richard Swedberg. Princeton, NJ: Princeton University Pres, 1994.

Premus, Robert. *Locations of High Technology Firms and Regional Economic Development.* Washington, DC: U.S. Government Printing Office, 1982.

PriceWaterhouseCoopers and NVCA. *MoneyTree Survey.* Q2-2007.

R

Rezendes, Diane, and E. Khirallah. "Silicon Valley and the Culture of More." *Information Week*, September 25, 2000.

Roberts, Jennifer. "Winning Workplaces." *San Jose Magazine*, November 2007.

Roberts, Timothy. "CEOs Optimistic on Hiring, Worried by Housing Costs, Traffic." *Silicon Valley/San Jose Business Journal*, May 30, 2007.

Rosenberg, David. *Cloning Silicon Valley: The Next Generation High-Tech Hotspots.* London: Pearson Education, 2002.

Roy, Sabir. "Not Quite the World's Silicon Valley." *ZDNet India News*, January 2, 2004.

S

Saxenian, AnnaLee. "The Urban Contradictions of Silicon Valley." *International Journal of Urban and Regional Research* 7, no. 2 (1983): 237-61.

Saxenian, AnnaLee. "Silicon Valley and Route 128: Regional Prototypes or Historic Exceptions?" In *High Technology, Space, and Society (Urban Affairs Annual Reviews)*, edited by M. Castells. Beverly Hills, CA: Sage, 1985.

Saxenian, AnnaLee. "In Search of Power: The Organization of Business Interests in Silicon Valley and Route 128." Working Paper, E53-470, Department of Political Science, MIT, Cambridge, Massachusetts, 1986.

Saxenian, AnnaLee. "The Political Economy of Industrial Adaptation in Silicon Valley." Ph.D. dissertation, Department of Political Science, MIT, 1988.

Saxenian, AnnaLee. "The Cheshire Cat's Grin: Innovation and Regional Development in England." *Technology Review* 71 (February-March 1988): 67-75.

Saxenian, AnnaLee. "In Search of Power: The Organization of Business Interests in Silicon Valley and Route 128." *Economy and Society* 18 (1989): 25-50.

Saxenian, AnnaLee. *Regional Regional Advantage: Culture and Competition in Silicon Valley and Route 128.* Cambridge, MA: Harvard University Press, 1996.

Saxenian, AnnaLee. "Regional Networks and the Resurgence of Silicon Valley." Working paper 508, Institute of Urban and Regional Development, University of California, Berkeley, 1989.

Saxenian, AnnaLee. "Regional Networks and Resurgence of Silicon Valley." *California Management Review* 33, no. 1 (1990): 89-112.

Saxenian, AnnaLee. *Silicon Valley's New Immigrant Entrepreneurs*. San Francisco: Public Policy Institute of California, 1999.

Saxenian, AnnaLee. *Local and Global Networks of Immigrant Professionals in Silicon Valley*. San Francisco: Public Policy Institute of California, 2002.

Saxenian, AnnaLee, and Chuen-yeah Li. "Bay-to-Bay Strategic Alliances: Network Linkages Between Taiwan and U.S. Venture Capital Industries." *International Journal of Technology Management* 25, nos. 1 and 2 (2003).

Saxenian, AnnaLee. "Government and Guanxi: The Chinese Software Industry in Transition." Discussion Paper, Center for New and Emerging Markets, London Business School, 2003.

Saxenian, AnnaLee, and Xiaohong Quan. "China." In *The Software Industry in Emerging Markets*, edited by Simon Commander. Northhampton, MA: Edward Elgar, 2005.

Saxenian, AnnaLee. *The New Argonauts: Regional Advantage in a Global Economy*. Cambridge, MA: Harvard University Press, 2006.

Schmidt, Eric, and Hal Varian. "Google: Ten Golden Rules," *Newsweek*. December 2, 2005.

Science Park Administration. *Science-Based Industrial Park*. Hsinchu, Taiwan, 1999.

Siegel, Lenny, and Herb Borock. "Background Report on Silicon Valley." *Report prepared for the U.S. Commission on Civil Rights*. Mountain View, CA: Pacific Studies Center, September 1982.

Siegel, Lenny, and John Markoff. *The High Cost of High Tech*. New York: Harper and Row, 1985.

Southwick, K. *High Noon: The Inside Story of Scott McNealy and the Rise of Sun Microsystems*. New York: John Wiley & Sons. 1999.

Southwick, Karen. "Java Takes Off." *Upside Today*, August 23, 1999.

Srinivasan, S. "Bangalor's Infrastructure Can't Keep Up with the Tech Boom." *Information Week*, September 2, 2003.

Stemlau, John. "Banglalore: India's Silicon Valley." *Monthly Labor Review* 119(2), November 1996, 50-52.

Sturgeon, Timothy J. "How Silicon Valley Came to Be." In *Understanding Silicon Valley: Anatomy of an Entrepreneurial Region*, edited by Martin Kenney. Stanford, CA: Stanford University Press, 2000.

Sturgeon, Timothy J. "The Origins of Silicon Valley: The Development of the Electronics Industry in the San Francisco Bay Area." Master's thesis, Department of Geography, University of California, Berkeley, 1992.

Suchman, Mark C., and Mia L. Cahill. "The Hired Gun as Facilitator: Lawyers and the Suppression of Business Disputes in Silicon Valley." *Law and Social Inquiry* 21 (1996): 679-712.

T

Tajna, Carolyn E. "Fred Terman—The Father of Silicon Valley." Report, Department of Computer Sciences, Stanford University, 1995.

Trajtenberg, Manuel. "Innovation in Israel 1968-1997: A Comparative Analysis Using Patent Date." *Research Policy* 30, no. 3 (2000): 363-389.

V

Von Hippel, Eric. *The Sources of Innovation*. New York: Oxford University Press, 1988.

W

Wadhwa, V., et. al. *America's New Immigrant Entrepreneurs*. Durham, NC: Duke University School of Engineering. January, 2007.

Williams, James C. "The Rise of Silicon Valley." Innovation and Technology, *California Management Review*, Spring/Summer, 1990.

Williams, Jeffrey R. "How Sustainable is Your Competitive Advantage?" *California Management Review*. Spring 1992, 29-51.

Wilson, John. *The New Venturers: Inside the High-Stakes World of Venture Capital.* Reading, MA: Addison-Wesley, 1985.

Wolfe, Tom. "The Tinkerings of Robert Noyce: How the Sun Rose on Silicon Valley." *Esquire Magazine,* December 1983, 346-74.

X

Xia, Yinqi. "China's Science Parks." Bridging the Digital Divide. Presentation by Deputy Director, Administrative Committee of Zhongguancun Science Park. University of California, Berkeley, April 2, 2004.

Xiao, Wei. "The New Economy and Venture Capital in China." *Perspectives* 3, no. 6 (September 30, 2002).

Y

Ye, Guo Bai. "Shanghai Silicon Intellectual Properties Transaction Center Established in China." *Xinhuanet.com,* October 3, 2003.

Yew, Lee Kuan. *From Third World to First: The Singapore Story, 1965-2000.* New York: Harper Collins, 2000.

Yoshida, Junko, and George Leopold. "Valley Entrepreneurs Look Homeward to China." *EE Times,* August 15, 2001.

Z

Zider, Robert. "How Venture Capital Works." *Harvard Business Review,* November-December 1998, 131-39.

Index

F

Facebook, 68, 117-118

Fairchild Eight, see Traitorous Eight

Fairchild Semiconductors, 27, 31-32, 80-81, 152, 169, 175-176

Farnsworth, Philo, 29

Federal renewable energy tax credit, 142

Federal Reserve Bank, 76

Federal Telegraph Company, 27, 29, 151

Filo, David, 33, 68

Financial angels, 75

Fisher Research Laboratories, 29

Flickr, 118

Folksonomy, 118

Fortune Magazine, 106, 156, 170

France, 97, 127, 134, 170

Fund managers, 80, 131

G

Gamow, David, 132

Gartner, Inc., 138

Gates, Bill and Melinda, 9, 14

General Electric, 30

Geothermal, 126

Germany, 84, 127, 133, 164

Global player, 67

Google, 27, 33, 68, 71, 80, 90, 92, 103, 122, 124, 126, 130, 136, 138-139, 141-142, 156, 181

Goss, Ernie, 111

Great Depression, 13, 149

Green technology, 43

Grinich, Victor, 32,

Grove, Andy, 135,174

Guardian angels, 74

Guardino, Carl, 100

H

Hardware, 32, 38-39

Hartford, Connecticut, 36

Harvard Business Review, 51, 172, 178-179, 184

Health care, 78, 100, 101-102

Hewlett-Packard, 27-28, 30, 103, 112, 130, 155

Hewlett, William, 29-30

High school education, 131-132, 160

Higher education, 101

Highspeed broadband, 117

Hoefler, Dan, 25, 151, 174

Hoerni, Jean, 32

Horizontal clustering, 37

Housing costs, 100-102, 129, 134

HP, see Hewlett-Packard

Hurley, Chad, 14

Hyper-competitive, 89

I

IBM, 30, 52, 106

Idea-based economy, 138

Immigrants, 71, 87, 106

India, 46, 61-64, 71, 87-89, 97, 127, 129, 147, 167-171, 180, 182

Indian community, 87

Industry cluster, 36-38, 57-58, 119

Information sector, 45, 47

Innovation ecosystems, 4, 147

Innovative regions, 171

Integrated circuits, 31, 88, 113, 168

Intel, 27, 32, 52, 60-61, 81, 90, 132, 135-136, 138-139, 158, 173-174

Intellectual property, 16, 30, 36, 52, 56, 58, 71, 84, 145, 148, 183

Internet, 33, 40, 42, 103, 114, 116-118, 124, 126, 128, 135, 138, 174

O

Obama administration, 93, 102

Old money, 76

Omidyar, Pierre, 33

Operational angels, 75

Oracle, 136

Outsourcing, 41, 43

P

Pacific Gas & Electric (PG&E), 15, 123-124, 127, 163

Pacific SolarTech, 124

Packard, David, 30

Page, Larry, 33, 68, 80, 124

Palm, Inc., 75

Palo Alto, 25-26, 32, 64, 71, 106, 124, 177

Pandora, 117

Parallada, Fransec, 161

Patents, 58-59, 62-65, 74, 111

Perk wars, 89

PG&E, Pacific Gas & Electric

Photovoltaic, 121, 125, 142

Pique, Josep, 161

Porter, Michael, 13, 37, 152, 179

Postrel, Virginia, 112, 156, 180

Productivity, 21, 37, 39, 46, 87-88, 95, 116, 157

Professional networks, 11, 20, 51, 53, 55, 68, 72, 89, 94-95, 97, 132, 145, 156,

Q

Qualcomm, 60-61, 63, 132

Quality of life, 11, 20, 54-55, 101, 104-107, 129, 132, 145, 147, 156

R

R&D, see Research and development

Raleigh-Durham, North Carolina, 31, 42

RCA, 29

Recession, 13, 35, 39-41, 113-114, 133, 137-139, 149, 159

Red de Espacios Tecnológicos de Anadalucia (RETA), 15

Regional economics, 14, 21, 28, 37, 39, 95, 103, 112, 161, 163, 80

Renewable, 120-121, 123, 126-128, 142, 159

Research and development (R&D), 60, 62, 64, 74, 136

Research Triangle, 14

Research universities, 11, 20, 26, 52, 55-65, 97, 116, 129-130, 145-148, 153

Resilience, 11, 15-16, 19, 54, 70, 111-115, 117, 119, 121,123, 125-127, 145, 156-157, 180

Resurgence, 15, 115, 127, 138, 152, 163-164, 179, 181

RETA, see Red de Espacios Technológicos de Anadalucia

Ripple effect, 29

Risks, 21, 28, 66-70, 76, 78, 129

Roberts, Sheldon, 32

Robinson, Joe, 133-134

Rockefeller, Laurence, 81

Romera, Felipe, 4, 161

Roscheisen, Martin, 124

Rosenfeld, Arthur, 14

Route 128, 14, 112, 172, 175-177, 180-181

Russia, 127

S

Sand Hill Road, 81, 124

SanDisk, 60-61, 91

SAP, 52, 71, 121, 138

Readers look to Nova Vista books for their excellent business, nature and music thinking and their appealing graphics. Our motto, "New Views that Work," inspires us to deliver proven, excellent ideas in every title. Take some time out and see!

PUBLISHING

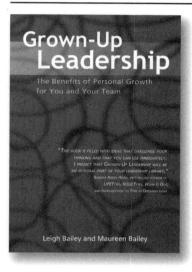

Grown-Up Leadership
The Benefits of Personal Development for You and Your Team

Leigh Bailey and Maureen Bailey

Why do leaders fail? Intimidators (who need to control) and Accommodators (who need acceptance) suffer from not knowing and accepting themselves. A holistic self-discovery process identifies your prism of background and experience and reveals their impact on your leadership style. Learning to use complementary styles, you will coach, motivate, and direct people more effectively. Practical advice, practices, tools, and stories in an inspiring, challenging book.

ISBN 90-77256-09-1. Paper, 144 pages, 160 X 230 mm (6 ½" X 9"), US$18.95

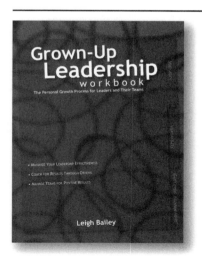

Grown-Up Leadership Workbook
The Personal Growth Process for Leaders and Their Teams

Leigh Bailey

This workbook helps leaders outgrow fears and limitations through a series of exercises, guided reflections, and self-assessments. A proven, holistic approach that has had a major impact on thousands of leaders' work and personal lives. You gain maturity, versatility and improved ability to coach and work productively with teams as you increase self-awareness and escape biases and blind spots. A free-standing coaching tool that leaders can use on their own to grow up fast.

ISBN 90-77256-15-6. Paper, 96 pages, 215 x 280 mm (8 ½ x 11"), US$14.95

To order, visit www.novavistapub.com

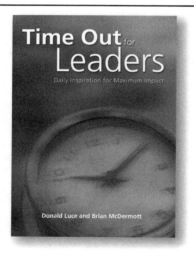